Biological Resource Management in Agriculture

CHALLENGES AND RISKS OF GENETICALLY ENGINEERED ORGANISMS

OECD

ORGANISATION FOR ECONOMIC CO-OPERATION AND DEVELOPMENT

ORGANISATION FOR ECONOMIC CO-OPERATION AND DEVELOPMENT

Pursuant to Article 1 of the Convention signed in Paris on 14th December 1960, and which came into force on 30th September 1961, the Organisation for Economic Co-operation and Development (OECD) shall promote policies designed:

- to achieve the highest sustainable economic growth and employment and a rising standard of living in member countries, while maintaining financial stability, and thus to contribute to the development of the world economy;
- to contribute to sound economic expansion in member as well as non-member countries in the process of economic development; and
- to contribute to the expansion of world trade on a multilateral, non-discriminatory basis in accordance with international obligations.

The original member countries of the OECD are Austria, Belgium, Canada, Denmark, France, Germany, Greece, Iceland, Ireland, Italy, Luxembourg, the Netherlands, Norway, Portugal, Spain, Sweden, Switzerland, Turkey, the United Kingdom and the United States. The following countries became members subsequently through accession at the dates indicated hereafter: Japan (28th April 1964), Finland (28th January 1969), Australia (7th June 1971), New Zealand (29th May 1973), Mexico (18th May 1994), the Czech Republic (21st December 1995), Hungary (7th May 1996), Poland (22nd November 1996), Korea (12th December 1996) and the Slovak Republic (14th December 2000). The Commission of the European Communities takes part in the work of the OECD (Article 13 of the OECD Convention).

PREFACE

These proceedings evolved from the OECD Co-operative Research Program workshop on "Challenges and Risks of GMOs – What Risk Analysis is Appropriate?" held in Maastricht, The Netherlands on 16-18 July 2003. The OECD Co-operative Research Program for "Biological Resource Management" has existed since 1990. It focuses on work in four specific areas, two of which are "New agricultural products for sustainable farming and industry" (Theme 1) and "Connecting scientific progress to sustainable and integrated agro-food systems" (Theme 4). These themes include priority areas such as transgenic organisms in agriculture and societal options and responses to ensure the availability and quality of resources. It proposes to widen the variety of goods offered by the agricultural producers, and to encourage and increase opportunities to manage engineered crops including raw materials for the fine chemical industry.

The other themes are "Quality of animal products and safety of food" and "Enhancing environmental quality in agricultural systems." The activities promoted by this program are post-doctoral fellowships (announced annually) and the organization of expert workshops (1-2 workshops per Theme per year). Currently participating countries in the program are; Australia, Austria, Belgium, Canada, Czech Republic, Denmark, Finland, France, Germany, Greece, Hungary, Ireland, Italy, Japan, Korea, The Netherlands, New Zealand, Norway, Poland, Portugal, Slovak Republic, Spain, Sweden, Switzerland, United Kingdom and the United States.

For further information about the OECD program, contact Mrs. Liliane Shettle, Head of the Program, Directorate for Food, Agriculture and Fisheries, OECD 2 rue Andre Pascal, F-75775 Paris Cedex France (email liliane.shettle@oecd.org, or consult the Internet site of the OECD (www.oecd.org/agr/prog/).

The workshop on "Challenges and Risks of GMOs - What Risk Analysis is Appropriate?" was motivated the fact that GMOs are facing a rather divergent public dispute all over the world with regards to environmental, safety and economic issues. This workshop complimented a training offered by the Amsterdam-Maastricht Summer University (AMSU) and the European Institute for Public Administration (EIPA) for agency representatives from various countries.

Presentations and discussions made it obvious that while the European Union now operates with one currency, many other types of policy decision remain to be worked out. Numerous examples illustrated the apparent dilemma between being politically correct in discussions between EU member countries while attempting to reach a consensus and still making progress on the issue at hand. The issue of coexistence between organic, conventional, and GM-based producers was thoroughly discussed.

Organic producers recently unsuccessfully petitioned to totally ban GM technologies from Northern Austria by suggesting the appropriateness of an artificially low level of GM contamination (<0.1%) in organic foods while organic standards permit the presence of up to 5% non-organically (or conventionally) grown products.

Surveys within the EU showed that many consumers equated GM technologies with the use of herbicides and pesticides and further they preferred these products not be used in food production. Participants agreed that credible consumer information is lacking in Europe. It was clear that EU consumers want GM products to be labelled so that consumers have a choice, but apparently the interest in food safety does not extend to other production practices that affect the environment (i.e., consumers only insist on GM labelling rather than including information about pesticide and fertilizer use).

Several EU companies are successfully using GM technologies for economic and environmental reasons in production of pharmaceuticals, but do not require labelling as GM products because the material contains no GM signature. Other GM-based food products that do not contain a GM signature are unknowingly being marketed in the EU. Lack of acceptance of GM products in the EU is apparently partially related to a perceived absence of benefits to consumers. Consumers need to be informed that reduced production costs associated with GM technologies only represent a very small portion of the cost of food.

These proceedings contains all papers presented during the workshop and available at the time of publication and summary of the discussion session. Papers and comments represent the different views regarding GM technology.

When possible speakers will update their contribution in relation with scientific, regulatory and other developments occurred since the time the workshop was held. The opinions and conclusions expressed in the workshop and these proceedings are those of the authors only and do not necessarily reflect the views of their organisations, nor of the workshop organisers, nor of the OECD and its member countries, nor of other countries concerned.

Maastricht, July 18 2003,

Professor Ervin Balázs
(Theme 1 Co-ordinator)
Agricultural Biotechnology Center
Environmental Biosafety Research Institute
H 2100 Gödöllö
Szent-Györgyi A street 4
Hungary

Professor James S. Schepers
(Theme 4 Co-ordinator)
Soil and Water Conservation Research USDA-ARS, NPA
University of Nebraska
113 Keim Hall
Lincoln, NE 68583-0915
USA

Acknowledgements

The Amsterdam-Maastricht Summer University (AMSU) and the European Institute of Public Administration (EIPA), organisers of this workshop, would like to express their gratitude to the following organisations and programmes for their financial support:

- The Co-operative Research Programme: Biological Resource Management for Sustainable Agricultural Systems of the Organisation for Economic Co-operation and Development (OECD),

- The Social Transformation Programme Central and Eastern Europe (Matra) of the Netherlands Ministry of Foreign Affairs and

- The Programme of Co-operation with CEE and NIS countries of the Netherlands Ministry of Spatial Planning, Housing and the Environment

The workshop organisers would also like to thank all participants and their staff for their valuable contribution towards a better understanding of the subject matter.

TABLE OF CONTENTS

PREFACE 3

SESSION 1   **Introduction**.. 9
             Bettina Rudloff
             The Concept of Risk Analysis................................ 11

SESSION 2   **The Multinational Dimension: Existing Legal Framework**............ 27
             Simonetta Zarilli
             International Trade in Biotechnology Products and Multilateral Legal
             Frameworks 29

SESSION 3   **Risk Analysis in Different Countries** 47
SESSION 3.1 **The European Union Dimension**.............................. 49
             Denise Prévost & Geert van Calster
             The EU Legislation regarding GMOs and its Implications for Trade ... 51
             Kim-Helleberg Madsen
             Food Safety: Novel Food, Labelling and Marketing of all Genetically
             Modified Feed and Food... 61
             Harry A. Kuiper
             Risk Analysis for GMOs and the Role of the New EFSA 63
             Jean-Luc Gal
             Intellectual Property Rights Regime in the EU: Directive 98/44 on the
             Legal Protection of Biotechnological Inventions 75
             Jeanine van de Wiel
             Biotech Food - Is It Safe Enough or is Safety Not Enough? A National
             Case Study from the Netherlands 87
SESSION 3.2 **State of Art in Other Countries**............................ 99
             Alan McHughen
             Agricultural GMOs: Risk Analysis and Intellectual Property Protection
             in the USA 101
             Masakazu Inaba & Darryl Macer
             Japanese Views on Biotechnology and Intellectual Property.............. 113
             Ruth Mackenzie
             Developing Countries and the Regulation of GMOs: Some
             Perspectives and Problems 127

SESSION 4   **Lessons Learned and Remaining Challenges**.................... 139
             Hubert P.J.M. Hubert Noteborn & Wim de Wit
             Scientific Challenges for Risk Assessment 141
             Enzo Gallori
             Risk Analysis of Soil-Plant Horizontal Gene Transfer 153

SESSION 5   **The Various Stakeholders' Positions** 159
                  Alexandra Hozzank
                  Sustainable Agricultural Systems and GMOs. Is Co-Existence
                  possible? 161
                  Piet van Dijck
                  The Processing Sector: Integration of Environmental Concerns in
                  Industrial Strategies 171
                  Sip de Vries
                  The Agricultural Sector: Impact on Agricultural Markets and
                  Competitiveness 181
                  Jerry Ploehn
                  A Farmer's Perspective on GMO Risk Analysis 191

                  Beate Kettlitz
                  The Consumer: The Right to Information and Specific Requirements
                  on Strategies 201
                  Katherine Williams
                  Challenges for the Media: Disseminating Information by Avoiding
                  Hysteria 203

SESSION 6   **General Discussion** 207

Annex 1   List of Participants 215
Annex 2   Organising Committee 223

SESSION 1:
INTRODUCTION

THE CONCEPT OF RISK ANALYSIS

Bettina Rudloff
European Institute of Public Administration EIPA
O.L. Vrouweplein 22
6201 BE Maastricht, The Netherlands

1. Background

Risk Analysis encompasses the relevant steps for handling food risks by differentiating the scientific level of Risk Assessment, the political level of Risk Management and the communicative level of Risk Communication.

The Codex Alimentarius Commission has developed guidelines for optimizing Risk Analysis by separating all levels functionally and especially by splitting Risk Assessment from Management.[1] This divorce is stressed to ensure the neutral basis for derivating independent, and objective management strategies.[2]

Additionally to these guidelines, the actual relevance of an independent Risk Assessment and a neutral Risk Management strategy is emphasised increasingly by current political developments:

- Food Standards have become a major barrier for international trade. More and more WTO disputes refer to the SPS-Agreement and focus on the question whether national Food Safety Standards are scientifically justified.[3]

- At the European level the establishment of the European Food Safety Authority (EFSA) has institutionally implemented the Codex guidelines to split Risk Management from Risk Assessment and thereby emphasises the importance of a neutral scientific Risk Assessment as part of its organisational statute.[4]

- Several countries reform their agricultural policies at present due to the ongoing WTO negotiations and domestic budgetary constraints for expensive agricultural support. As a consequence a shift from WTO-incompatible measures (like traditional income support instruments) towards compatible

---

[1] See for example Codex Alimentarius (2003): Report of the Eighteenth Session of the Codex Committee on General Principles, Paris, France, 7-11 April 2003, ALINORM 03/33A, para. 10-31.

[2] See Codex Alimentarius Commission (2003): Report of the Twenty-sixth Session, FAO Headquarters, Rome, 30 June – 7 July 2003, ALINORM 03/41.

[3] See WTO: Index of disputes issues, available at http://www.wto.org/english/tratop_e/dispu_e/ dispu_subjects_index_e.htm#bkmk1, state December 2003.

[4] See EU (2002): Regulation (EC) No 178/2002 of the European Parliament and of the Council of 28 January 2002 laying down the general principles and requirements of food law, establishing the European Food Safety Authority and laying down procedures in matters of food safety, Official Journal of the European Communities L 31/1, Art. 18.

measures (like service-related subsidies, e.g. for ensuring Food Quality) can be noticed in a lot of countries.[5]

Despite the unquestioned general consensus on the increasing role of a sound Risk Analysis to tackle Food Safety issues there is still need to specify content, problems or requirements for each single level and the interlinkages between the levels (Figure 1):

Figure 1: Needs for Clarification of Risk Analysis

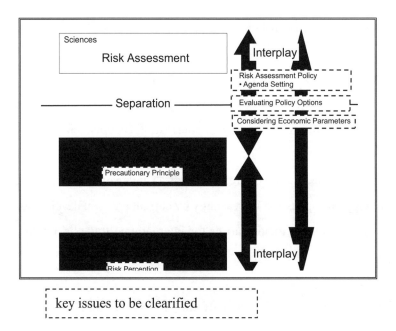

The first issues to be clarified in detail are referring to the interplay of the different levels and their best separation: i.e. at what level to integrate Risk Assessment Policy best, where to evaluate policy options and at what stage to consider economic parameters.

Additional issues are the appropriate consideration of public risk perception and the specification of the Precautionary Principle as concrete and operational management strategy.

These questions will be addressed in this paper by contestualyizing them to the specific characteristics of GMOs. Starting point for the Analysis will be the political level of Risk Management and the identification of what are the policy's demands on science.

---

[5] See OECD (2002): Agricultural Policies in OECD Countries: A Positive Reform Agenda,2002. Available at: www.oecd.org/agr.

2. Scope and Limitations for an Appropriate Risk Analysis for GMOs

2.1 What is Management Demanding on Risk Assessment?

Risk Managers do have specific information needs related to Risk Assessment, namely the provision of understandable and unambiguous information that easily can be translated into Management strategies.[6]

Figure 2: Usability of scientific Risk Assessment Results for Risk Management

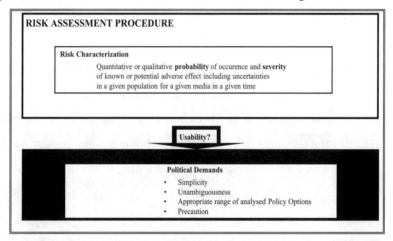

Codex Alimentarius Commission (2003): Report of the Twenty-sixth Session, FAO Headquarters, Rome, 30 June – 7 July 2003, ALINORM 03/41, Annex 1.

The required information from a political point of view can conflict with the results of a Risk Assessment procedure.

The political needs to derivate a Management Strategy can be summarized as follows:[7]

(1) Simplicity: the final information should be simple and understandable.

(2) Unambiguousness: different or contradictory information should be avoided.

(3) Policy Options: a sufficient but cost-extensive range of policy options should be covered.

(4) Precaution: an operational concept for implementing a political precautionary strategy should be developed.

---

[6] See The Presidential/ Congressional Commission on Risk Assessment and Management (1997): Risk Assessment and Risk Management in Regulatory Decision-Making, Final Report Volume 2, Washington D.C.

[7] Ibid., p.7

2.2 How May Risk Assessment Fulfil Management's Demands?

The mentioned demands will be applied to GMOs in the following.

(1) The requirement of simplicity is facing a general problem of science, i.e. *uncertainty* which may be especially high for so-called "new risks" like GMOs, where no sufficient empirical historical data can serve as base for derivating probabilities and severity of hazards. As this problem is a systematic one it is often recommended to be as transparent as possible on the limitations of scientific results.[8]

Additionally, new and food-related risks like GMOs are facing the problems of long-term and synergetic effects which are always difficult to assess.[9] A further specific problem for GMOs concerns the complexity of interrelated effects within ecosystems.[10]

Economic approaches may support the understanding of results and make them usable for politicians by targeting at directly relevant information for weighing management options.[11]

Typical economic elements to be considered in a benefit-cost-analysis are illustrated within the following table which simply indicates the potential effects. Hereby only the general impact independently from its negative or positive consequences and independently from its extent is mentioned.

---

[8] Regulation (EC) No 178/2002 …, Art. 38.
[9] See International Council for Science (2003): New Genetics Food and Agriculture: Scientific Discoveries – Societal Dilemmas, Doyle Foundation, Glasgow, p. 23.
[10] Ibid, p. 31.
[11] The Presidential/ Congressional Commission on Risk Assessment and Management (1997): Risk Assessment … Vol.1, p. 93-101.

Figure 3: Potential Economic Effects of GMOs

Area of potential effects	Potential impact on …
Producer Effects	• … productivity: yields, inputs • … input costs • … output prices
Consumer Effects	• … consumers' surplus by changed final food prices • … utility due to changed product quality (e.g. durability, nutritional value) • … utility due to asymmetric information (not individually controllable)
Sustainability	• … pesticide input • … toxigens of soil • … abiotic stress resistance • … gene transfer and biodiversity • … food security
Health	• … utility or externalities due to changed nutritional quality • … toxicity, allergencity, antibiotic resistance
Research	• … benefit allocation via innovative techniques, patents
Ethics	• … perception regarding equity and general limitations of science

International Council for Science (2003): New Genetics, Food and Agriculture: Scientific Discoveries – Societal Dilemmas, Doyle Foundation, Glasgow.

A general advantage of such cost-benefit-analyses is the possibility of summarizing several elements by aggregating and monetizing the different impacts. Additionally, the distribution of negative and positive effects among different addressed stakeholders can be identified. This might serve to weighten and justify policy options.

To ensure an interdisciplinary approach, where scientists may deliver the base information to Economists, a close co-operative work is needed. Such information required for economic analyses is e.g.:

- the distribution of damage probabilities to calculate the deviation of expected values and variances

In addition to general scientific uncertainties there exist specific weaknesses of economic methods such as quantification or monetization of effects and the determination of Consumers' utility.[12]

---

[12] International Council for Science (2003): New Genetics, Food and Agriculture: Scientific Discoveries – Societal Dilemmas, Doyle Foundation, Glasgow.

Regarding economic methods, a guideline where economic information should be integrated best does not exist. Whereas the SPS-Agreement proposes the integration of an economic evaluation into Risk Assessment, the Codex Alimentarius Commission defines economic parameters as part of Management. For transparency reasons a common terminology should be targeted.

Figure 4: Integration of Economic Factors within Risk Assessment

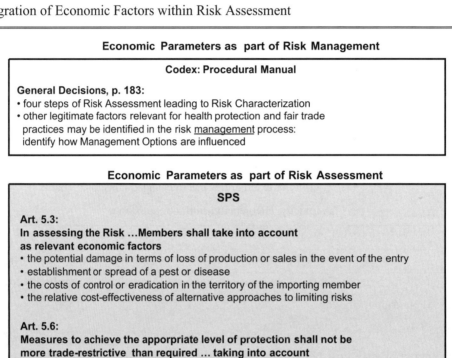

WTO (1994): Agreement on the Application of Sanitary and Phytosanitary Measures and Codex Alimentarius Commission (2003): Procedural Manual, Codex Statements of Principle Concerning the Role of Science in the Codex Decision-Making Process and the Extent to which other Factors are taken into account, Appendix: General decisions, 13th Edition.

(2) The avoidance of contradictory results is a dilemma for scientific work that is dynamic, developing and thereby may provide changing and different results. The institutional implementation can be different as some arrangements are explicitly aiming at unambiguous results[13] whereas others accept the general necessity of different results.[14]

(3) Risk Assessment Policy means the evaluation of different policy options concerning their potential to reduce a risk and should facilitate a Manager's decision. According to the Codex Principles it is recommended to be part of Risk Management.[15] However, actual institutional implementations show that often political recommendations are already been made at the level of Risk Assessment, such as in one of the latest briefings of the EFSA on Semicarbazide (SEM).[16]

(4) Even if the relevance and necessity of the Precautionary Principle is claimed politically its concrete specification is causing problems.[17] There are different operational interpretations depending on the level of Risk Analysis where to consider precaution:

- as part of Risk Assessment it may be integrated by safety margins related to probabilities or by establishing neutral criteria for defining non-negligible risks
- as part of Management it would be handled as political decision on time and levels of interventions to reduce risks.[18]

Several approaches on how to implement this principle can be found looking at different arrangements. The SPS-Agreement has established the possibility to introduce provisional precaution measures while further scientific assessment has to be undertaken for final measures.[19] This stresses the relevance of scientific evidence and simply postpones the final justification which is still necessary for stricter standards than the Codex ones.

While emphasising the general need for Precaution, the European Commission sets out guidelines for limiting its arbitrary use which, among others, covers the guarantee of the proportionality of the measure

---

[13] See Regulation (EC) No 178/2002 ... Art. 47.

[14] See WTO (1998): EC Measures Concerning Meat and Meat Products (Hormones), Report of the Appellate Body, WT/DS26/AB/R, 16. January 1998, para. 510.

[15] See Codex Alimentarius (2003): Report of the Eighteenth Session ..., para. 16.

[16] In addition to the assessment that risks to consumer were very small, the EFSA published the advice to reduce SEM in baby foods as swiftly as technological progress allows. See EFSA (2003): Advice on semicarbazide, in particular related to baby food, Ad hoc expert group meeting, 9 October 2003, EFSA/AFC/adhoc SEM/2-, available at http://www.efsa.eu.int/pdf/pressrel20031015_en.pdf.

[17] Woolcock, S. (2002): The Precautionary Principle in the European Union and its impact on international trade relations, CEPS working document No. 186.

[18] See International Council for Science (2003): New Genetics ... p. 20.

[19] WTO (1994): Agreement on the Application of Sanitary and Phytosanitary Measures. Article 5.7.

or the choice of the less restrictive intervention. Thereby the European approach does not focus on methodological and universal criteria for precaution but rather on implementation issues.[20]

Summarizing the described problems for Risk Assessment, no unambiguousness upon which to base political decision can be identified.

If scientific Risk Assessment cannot provide non-ambiguous information for defining political options additional rules for basing Risk Management should be considered what will be discussed in the following.

2.3 What Role Plays Communication for Risk Management?

Two major objectives of Risk Communication can be differentiated into:

(1) the information and education of the public, i.e. improving the understanding of risk, correcting perceptual biases, explaining risk to lay people

and

(2) the handling with conflicting views and interests to avoid conflict escalations in disputes about risks and to minimise differences in evaluation.[21]

Communication is not an isolated element but should refer to all levels of Risk Analysis.

When related to the first informative aim, communication is needed at the level of Risk Assessment among the different assessors with different expertise. Additionally it is needed between scientists and managers to understand limitations of scientific results and political requirements on information.

Finally the communication to different public groups is of high relevance for the acceptance and trust into management strategies.

Whereas scientific Risk Assessment is based on the quantitative coverage of probabilities and damage amounts (Figure 2) the public risk perception also considers qualitative parameters of lay persons.

---

[20] European Commission (2000): Communication from the European Commission on the Precautionary Principle.

[21] Wiedemann, P. M.; Schütz, H. (2000) : Developing Dialogue-Based Communication Programmes. Arbeiten zur Risiko-Kommunikation, Heft 79. Programmgruppe Mensch, Umwelt, Technik des Forschungszentrums Jülich, Jülich. Available at *http://www.fz-juelich.de/mut/hefte/heft_79.pdf*

Figure 5: Parameter influencing the Risk Tolerance

Dimensions influencing Public Perception	Risk Tolerance
Individual observation	
Individual control	
Equal distribution of costs and benefits	
Dread	
Artificiality	
Delay of Effect	

Source: Renn, O., Rohrmann, B. (1999): Risk Perception Research - An Introduction, p. 31, in: Renn, O., Rohrmann, B. (ed.): Cross-Cultural Risk Perception. A survey of Empirical Studies, Kluwer Academic Publishers, Dordrecht-Bostin-London, p. 31.

Applied to GMOs a low "Risk Tolerance" can be concluded looking at the relevant parameters of missing individual observability and control and by the perceived allocation of positive (industrial companies, conventional farmers) and negative consequences (the public). In addition, perceived dread, artificiality and an expected delay of the damage decrease Risk Tolerance.[22]

Risk tolerance can differ among countries. This causes diverging optimal national management strategies leading to potential trade conflicts.[23]

On the basis of the perception criteria and according to cross-national surveys the following conclusions for politicians can be drawn:[24]

(1) Open communication of restricted science

Generally science's limits are accepted by the public and should be communicated in a more open way to build up a common consensus on the general lack of full control. Not to deny these systems immanent limits may support the trust in institutions. This is especially relevant for risks characterised by low perceived knowledge (new risks) and dread (involuntary risks).

(2) Acceptance of ethical concerns as relevant part of risk assessment

The need for considering broader and general ethical questions is expressed as especially urgent when no evidence of risk is stated. Again, trust in institutions may be supported by accepting and communicating these ethical issues not as individual and irrational concern but as a societal task.

---

[22] See Renn, O., Rohrmann, B. (1999): Risk Perception Research ... p. 31.
[23] Ibid.

(3) Choosing correct means of public participation

Often the stronger involvement of the public is stressed as a standard and simplified way for risk communication. However, more participation does not lead automatically to more legitimacy and democracy. The understanding of knowledge should be broadened to consider social strength, i.e. the ability to respond flexibly to political, social and economic needs, in addition to the pure expert's knowledge. Therefore, real and equitable considerations as well as the analysis of relevant actors including layers' perceptions should be targeted instead of a complex set of formal participatory rights.

(4) Important further research on perception[25]

Further empirical research may increase the understanding of perception and the influencing factors. Especially the following influencing criteria may be relevant to be examined:

- Cross-cultural Studies to gather differences in risk perceptions due to different traditions and cultures. This is of increasingly relevance within growing trade.

- Demographic factors as gender and age to understand the development of risk perception.

- The impact of individual experience with a specific risk or hazard on the respective risk perception.

- The importance of individual risk attitudes such as sensation seeking or risk propensity.

- The link between actual behaviour and expressed judgement, i.e. the transfer of statements into real action like boycotts.

Taking into account the public risk perception should not be understood in an isolated way but be embedded within the overall Risk Analysis package. As part of Risk Management two major benefits can be identified:

(1) The *choice of a certain instrument* can depend on the underlying perception parameters: related to GMOs such instruments are those influencing the low controllability and observability (see fig 6). For instance, a label may increase the risk tolerance.

(2) The perceived risk and the felt restriction to individual protection are influencing public *trust in institutions*. Especially when the availability of data and the individual options for protection is perceived by the public as being limited, the need for credibility may increase. Therefore, performance and trust in responsible institutions is of high relevance for GMOs.[26]

---

[24] See Marris, C.; Wynne, B.; Simmons, P.; Weldon, S. (2001): Public Perceptions of Agricultural Biotechnologies in Europe. Final Report of the PABE research project funded by the European Communities, FAIR CT98-3844 (DG12-SSMI).

[25] Renn, O., Rohrmann, B. (1999): Cross-Cultural Risk …

[26] Slovic, P. (2000): Perceived risk, trust and democracy, in: Slovic, P.: Risk Perception. London and Sterling, S.316-326

As science cannot provide non-ambiguous guidelines for risk management, the integration of stakeholders into all the steps of risk analysis can increase the acceptance of chosen risk management strategies.

2.4 Case Study on a Communicative Measure as Management Instrument: Labelling as Silver Bullet?

Labelling schemes often are expressed as silver bullet for Food Safety Policy or consumer protection.

A first advantage of a label often stressed is the guarantee of the consumers' sovereignty by leaving the freedom of choice for certain products. Thereby economic efficiency and the internationally liberal trade can be achieved.[27]

Another benefit mentioned refers to traceability, i.e. the long-term possibility of monitoring food as well potential risks after approval and marketing.[28]

The success of a label may be hindered if the labelled product attributes such as "containing GMOs" or "produced without GMOs" cannot be controlled individually.

Figure 6: Pros and Cons of a Label Scheme

Pros	Cons
Supporting consumers' sovereignty and liberal trade	Asymmetric Information on underlying credence attributes leading to:
Replacing Missing individual observation	
Supporting post-market surveillance	• risk of misuse
	• loosing trust into label information

The visibility of product attributes to the consumer influences the effectiveness of management instruments. Credence attributes cannot be physically experienced and product differentiations such as "GMO-containing" and "GMO-free" cannot be controlled by the consumer. This phenomenon is known as "Lemon Market" and is describing one reason for market failures which justifies political intervention into a free market.[29]

Consider two market segments, one low-cost conventional food production fulfilling lax GMO standards and another high-cost organic production requiring stricter GMO residua standards.[30]

Due to different consumer preferences[31] it is assumed that the organic food is reaching higher prices.

[2727] Stressed as instrument achieving both, health protection and promoting fair trade practices. See Codex Alimentarius Commission (2003): Procedural Manual …para. 3.

[28] International Council for Science (2003): New Genetics … , p. 25.

[29] Akerlof, G.A. (1970): The Market for "Lemons": Uncertainty and the Market Mechanism, in: Quarterly Journal of Economics 84, S. 488 – 500.

[30] This problems of Co-Existence and he impact on the organic sector is currently being discussed at the European level in the context of the new Regulation on Labelling and Traceability.

In addition to the pure technological difficulty of ensuring Co-existence, the missing differentiation by the consumer may cause problems: the compliance of laxer or stricter GMO-requirements is just known by the producer but not detectable by the consumer[32].

This lack of controllability causes the fear of a misuse of the Label on the "strict GMO-standard" to reach the highest realizable prices and assets. Due to this insecurity on the actual product quality, consumers may reduce their willingness to pay the higher prices. The resulting price pressure can cause a destruction of the organic food sector if the higher production costs cannot be covered anymore by the lower prices.

Once again trust in institutions is of high relevance and can be supported by reliable guidelines for Co-existence[33] and strong monitoring and controlling procedures for label, e.g. by establishing reliable penalties or by strong private brands.[34]

3. Perspectives for Future Risk Analysis

Several developments influence the future Risk Analysis at each single level.

With regard to *Risk Assessment* the following topics may be characterized:

- increased problems for the so-called second generation of GMOs: So far rather abiotic characteristics like internal insecticides are addressed whereas changed nutritional aspects are expected in the future.[35] Scientific uncertainties may increase when attributes are getting more complex. On the contrary the perceived risk for these attributes may be lower as the consumer is already used to nutritional changes such as related to functional food.

- Recently the Codex Alimentarius is developing Risk Analysis Guidelines for GMOs which will strengthen the binding character of Risk Assessment at the multinational level as it is foreseen in the SPS-Agreement.[36]

---

[31] OECD (1999): Food Safety and Quality Issues – Trade Considerations, OECD's consultant's report by Bureau, J.C., Marlette, S., Gozlan, E. and Jones, W., published under the responsibility of the Secretary General, Paris, p. 20.

[32] Known as asymmetric information on product attributes.

[33] EU Commission (2003): Commission Recommendations of July 2003 on guidelines for the development of national strategies and best practices to ensure the co-existence of genetically modified crops with conventional and organic farming.

[34] Label Schemes are a key issue within the new European Legislative Package on GMOs and are meant to be applied as well to highly refined oils without any detectable DNA . The problem of physical traceability may become a problem at WTO level: the WTO traditionally follows a strict separation of physically detectable standards and standards without physical impact. See EU Commission (2003): State of play on GMO authorisations under EU law, Press Release 7 November 2003, MEMO/03/221.

[35] See International Council for Science (2003): New Genetics …

[36] See Codex Alimentarius Commission (2003): Report of the Twenty-sixth Session …, par. 230.

With regard to *Risk Management* the potentially developed Standards at the Codex Alimentarius will reduce the scope for national standards and for applying those to imports.

Finally, the recommended split of Risk Assessment and Management has been applied in different countries in various ways and must not necessarily be the only successful way.

Following the structure of the new European Food Safety Authority the majority of European Countries implement this separation. Nevertheless, there are successful examples of countries without that split such as the UK.[37] Further, even if neutrality and objectivity of Management is strongly expressed from the US at the level of WTO negotiations, the Food and Drug Administration, as the responsible national body, is in charge of both tasks.[38]

References

(1) Akerlof, G.A. (1970): The Market for "Lemons": Uncertainty and the Market Mechanism, in: Quarterly Journal of Economics 84, p. 488 – 500.

(2) Barth, R., Brauner, R., Hermann, A. et al. (2003): Genetic Engineering and Organic Farming, Federal Environmental Agency, Report 02/03.

(3) Bureau, J.-C., Marette, S., Schiavana, S. (1998): Non-tariff barriers and consumers' information: the case of the EU-US trade dispute over beef, in: European Review of Agricultural Economics, 4/98, p. 437-462.

(4) Codex Alimentarius (2003): Report of the Eighteenth Session of the Codex Committee on General Principles, Paris, France, 7-11 April 2003, ALINORM 03/33A.

(5) Codex Alimentarius Commission (2003): Procedural Manual, Codex Statements of Principle Concerning the Role of Science in the Codex Decision-Making Process and the Extent to which other Factors are taken into account, Appendix: General decisions, 13th Edition.

(6) Codex Alimentarius Commission (2003): Report of the Twenty-sixth Session, FAO Headquarters, Rome, 30 June – 7 July 2003, ALINORM 03/41.

(7) EFSA (2003): Advice on semicarbazide, in particular related to baby food, Ad hoc expert group meeting, 9 October 2003, EFSA/AFC/adhoc SEM/2-. Available at http://www.efsa.eu.int/pdf/pressrel20031015_en.pdf.

(8) EU Commission (2003): State of play on GMO authorisations under EU law, Press Release 7 November 2003, MEMO/03/221.

(9) International Council for Science (2003): New Genetics, Food and Agriculture: Scientific Discoveries – Societal Dilemmas, Doyle Foundation, Glasgow.

---

[37] See OECD (2000): Compendium of National Food Safety Systems and Activities, Ad Hoc Group on Food Safety, SG/ADHOC/FS(2000)5/ANN/FINAL, 15 May 200, Paris.
[38] Ibid.

(10) Marris, C.; Wynne, B.; Simmons, P.; Weldon, S. (2001): Public Perceptions of Agricultural Biotechnologies in Europe. Final Report of the PABE research project funded by the European Communities, FAIR CT98-3844 (DG12-SSMI).

(11) OECD (1997): Uses of Food Labelling Regulations, OECD Documents, OECD/GD(97)150, Paris.

(12) OECD (1999): Food Safety and Quality Issues – Trade Considerations, OECD's consultant's report by Bureau, J.C., Marette, S., Gozlan, E. and Jones, W., published under the responsibility of the Secretary General, Paris.

(13) OECD (2000): Compendium of National Food Safety Systems and Activities, Ad Hoc Group on Food Safety, SG/ADHOC/FS (2000)5/ANN/FINAL, 15 May 200, Paris.

(14) Renn, O., Rohrmann, B. (1999): Cross-Cultural Risk Perception: State and Challenges, p. 227-228, in: Renn, O., Rohrmann, B. (eds.): Cross-Cultural Risk Perception. A survey of Empirical Studies, Kluwer Academic Publishers, Dordrecht-Bostin-London.

(15) Slovic, Paul: Perceived risk, trust and democracy: In: Slovic, P.: Risk Perception. London and Sterling, 2000, S.316-326.

(16) The Presidential/ Congressional Commission on Risk Assessment and Management (1997): Framework for Environmental Health Risk Management, Final Report Volume 1, Washington D.C.

(17) The Presidential/ Congressional Commission on Risk Assessment and Management (1997): Risk Assessment and Risk Management in Regulatory Decision-Making, Final Report Volume 2, Washington D.C.

(18) Wiedemann, P. M.; Schütz, H. (2000): Developing Dialogue-Based Communication Programmes. Arbeiten zur Risiko-Kommunikation, Heft 79. Programmgruppe Mensch, Umwelt, Technik des Forschungszentrums Jülich, Jülich. Available at *http://www.fz-juelich.de/mut/hefte/heft_79.pdf*

(19) Woolcock, S. (2002): The Precautionary Principle in the European Union and its impact on international trade relations, CEPS working document No. 186.

(20) WTO: Index of disputes issues, available at http://www.wto.org/english/tratop_e/dispu_e/ dispu_subjects_index_e.htm#bkmk1, State December 2003.

Agreements, Legal Acts, Communications

(1) EU Commission (2003): Commission Recommendations of July 2003 on guidelines for the development of national strategies and best practices to ensure the co-existence of genetically modified crops with conventional and organic farming.

(2) EU Commission (2003): Regulation (EC) No 1830/2003 of the European Parliament and of the Council of 22 September 2003 concerning the traceability and labelling of genetically modified organisms and the traceability of food and feed products produced from genetically modified organisms and amending Directive 2001/18/EC, Official Journal of the European Communities L 268/24.

(3) EU (2002): Regulation (EC) No 178/2002 of the European Parliament and of the Council of 28 January 2002 laying down the general principles and requirements of food law, establishing the European Food Safety Authority and laying down procedures in matters of food safety, Official Journal of the European Communities L 31/1.

(4) EU Commission (2000): Communication from the European Commission on the Precautionary Principle. COM(2000) 1

(5) WTO (1994): Agreement on the Application of Sanitary and Phytosanitary Measures.

(6) WTO (1998): EC Measures Concerning Meat and Meat Products (Hormones), Report of the Appellate Body, WT/DS26/AB/R, 16. Januar 1998.

SESSION 2:
THE MULTINATIONAL DIMENSION:
EXISTING LEGAL FRAMEWORK

INTERNATIONAL TRADE IN BIOTECHNOLOGY PRODUCTS AND MULTILATERAL LEGAL FRAMEWORKS

Simonetta Zarilli[*]

Division on International Trade and Commodities, UNCTAD Secretariat.
Palais des Nations, 8-14, Av. de la Paix
1211 Geneva. Switzerland

1. Introduction

Biotechnology is a revolutionary technology.[2] It offers humanity the power to change the characteristics of living organisms by transferring the genetic information from one organism, across species boundaries, into another organism. These solutions continue the tradition of selection and improvement of cultivated crops and livestock developed over the centuries. However, biotechnology identifies desirable traits more quickly and accurately than conventional plant and livestock breeding and allows gene transfers impossible with traditional breeding. The use of biotechnology in sectors such as agriculture and medicine has produced a growing number of genetically modified organisms (GMOs) and products derived from them. Changing the characteristics of organisms may provide benefits to society, including new drugs and enhanced plant varieties and food. However, biotechnology does not come without risks and uncertainty. Its potential effects on the environment, human health and food security are being actively debated at the national and international levels. Countries' positions depend on many factors, such as their policy awareness, the level of risk they are willing to accept, their capacity to carry out risk assessments in the sector and implement adequate legislation, their perception of the benefits they could gain from biotechnology, their dependence on agricultural exports, and the investments they have already made in the sector. However, there is a sharp contrast at present between the widespread international acceptance of biotechnology's benefits in pharmaceuticals and industrial products, and the widespread concerns about its possible dangers in agricultural and food production.

The global area of genetically modified crop plantation has grown 35-fold since 1996, and the current estimated global GM crop area is around 60 million hectares, grown by almost six million farmers in sixteen countries. Herbicide tolerant soybean is the dominant transgenic crop followed by Bt maize and herbicide tolerant canola. Four counties account for 99 per cent of the global transgenic crop area (United States, 66 per cent of global total; Argentina, 23 per cent; Canada, 6 per cent; and China 4 per cent).

---

[1]Legal Officer, Trade Negotiations and Commercial Diplomacy Branch, Division on International Trade and Commodities, UNCTAD Secretariat, Geneva, Switzerland.The author is grateful to I. Musselli for her useful comments. All mistakes are the author's only. The views expressed in this article are those of the author and should not be construed as the official UNCTAD position on the issues concerned. This paper draws on two previous papers by the author on related subjects: International Trade in Genetically Modified Organisms and Multilateral Negotiations – A New Dilemma for Developing Countries, UNCTAD/DITC/TNCD/1, 20 October 2000; and "Non-Trade Concerns and the WTO Jurisprudence in the Asbestos Case – Possible Relevance for International Trade in Genetically Modified Organisms" (in collaboration with I. Musselli), The Journal of World Intellectual Property, Vol. 5, No. 3, May 2002.

[2] The Convention on Biological Diversity defines biotechnology as "any technological application that uses biological systems, living organisms, or derivatives thereof, to make or modify products or processes for specific use". The biotechnology industry provides products for human health care, industrial processing, environmental bioremediation, and food and agriculture.

Minor plantings can be found in South Africa, Australia, Romania, Mexico, Bulgaria, Spain, Germany, Uruguay, Indonesia and Colombia.[3]

The diverging country appreciations of the risks and benefits related to agro-biotechnology have led to equally diverging domestic regulations on GM approval, labelling, documentation requirements, etc. When GM products are commercialized internationally, as it has been the case since the second half of the 1990s, the different domestic requirements may hamper international trade and further complicate an already difficult regulatory trade system in agricultural products. The rate of technological advance in biotechnology, on the other hand, is likely to be very rapid, meaning that the commercial life of any new GMO is likely to be short. This means that easy and quick access to foreign markets is a critical determinant for profitability.[4]

At present, international trade in GMOs has to take place according to the rules agreed by WTO Members at the end of the Uruguay Round, in particular those spelt out in the Agreement on the Application of Sanitary and Phytosanitary Measures (SPS Agreement), the Agreement on Technical Barriers to Trade (TBT Agreements) and in the General Agreement on Tariffs and Trade (GATT) 1994. Being biotechnology by and large a proprietary technology, the rules of the WTO Agreement on Trade-Related Aspects of Intellectual Property Rights (TRIPS) may also have a bearing on international trade in GMOs. Disciplines regarding transboundary movement of GMOs, however, have also emerged from specific multilateral agreements being negotiated outside the purely trade context, in particular the Cartagena Protocol on Biosafety. The rules included in different legal instruments may not be fully consistent with each other and may give rise to conflicts between GMO-exporting countries and potential importers.

This paper is intended to analyze certain relevant features of the Biosafety Protocol and of the above-mentioned WTO Agreements, assess potential substantive conflicts and tensions between WTO law and the Protocol itself, and explore how the two legal frameworks could apply and be possibly reconciled in the case of a trade conflict involving GMOs or GM products. The paper only focuses on agricultural biotechnology.

2. The Cartagena Protocol on Biosafety

This section of the paper is not intended to analyze the Biosafety Protocol in detail, but to single out those trade-related aspects of it that exhibit the greater potential for tension with WTO law.[5]

The Cartagena Protocol on Biosafety,[6] which was negotiated under the auspices of the Convention on Biological Diversity (CBD, Rio de Janeiro, 1992), was adopted on 29 January 2000, after almost four years of increasingly complex negotiations. It entered into force on 11 September 2003, 90 days after the 50th instrument of ratification was received. As of 1 November 2003, 66 countries, including the EC, had ratified it. The first Conference of the Parties meeting is scheduled to be held in Malaysia in February 2004.

---

[3] C. James, "Report: GM Crop Update 2002", International Service for the Acquisition of Agri-biotech Applications (ISAAA), available at: www.isaaa.org .

[4] Phillips P.W.B. and W.A. Kerr, "Alternative Paradigms – The WTO Versus the Biosafety Protocol for Trade in Genetically Modified Organisms", *Journal of World Trade*, 34(4), 2000, pp. 63 ff.

[5] For a detailed and comprehensive description and analysis of the Biosafety Protocol see: *An Explanatory Guide to the Cartagena Protocol on Biosafety*, IUCN Environmental Policy and Law Paper No.46, 2003.

[6] In general, the term "biosafety" describes a set of measures used to assess and manage any risk associated with GMOs.

Negotiating the Protocol in the framework of the CBD made it a predominantly environmental agreement. The environmental ministers took the leading role during the negotiations, as opposed to the trade ministers who negotiated the WTO Agreements and are currently involved in carrying out the Doha Work Programme.[7] This specific framework may explain why the large majority of developing countries took some negotiating positions that have been constantly rejected within the WTO context, such as those on the precautionary principle; secondly, it provides the ground for the Protocol to be conceived and still be an instrument primarily concerned with the conservation and sustainable use of biological diversity, more than with international trade *per se*.

The Protocol enters into force at a critical time in trade policy, with increasing tensions about the restrictive trade regime applied by certain countries on agro-biotechnology, and at a moment where, after the failure of the 5th WTO Ministerial Conference, the whole multilateral trade system seems to be at risk.[8]

The Protocol provides rules for the safe transfer, handling, use and disposal of "living modified organisms" (LMOs).[9] Its aim is to address the threats posed by LMOs to biological diversity, also taking into account risks to human health. Living modified organisms are defined by the Protocol as "any living organism that possesses a novel combination of genetic material obtained through the use of modern biotechnology" (Article 3(g)). The Protocol distinguishes LMOs in two categories: LMOs for voluntary introduction into the environment –such as seeds for planting, live fish for release, micro-organisms for bioremediation; and LMOs intended for direct use as food or feed, or for processing (LMO-FFPs). The latter includes the large majority of LMOs: genetically modified commodities, such as soybean, maize, canola, tomato, cotton, etc. The Protocol does not cover consumer products derived from LMOs, such as corn flakes, flour, starch, tomato paste or ketchup.

It seems there are three aspects of the Protocol which may specifically give rise to overlaps and tensions with WTO law: i. the scope for legitimate government action short of conclusive scientific evidence; ii. risk assessment and risk management; iii. and the socio-economic factors which may be taken into account in the decision-making process.

The Protocol permits the countries of import to take a precautionary approach;[10] this means that lack of scientific certainty due to insufficient information on the potential effects of LMOs on biodiversity, taking also into account risks for human health, will not prevent the receiving country from taking decisions regarding shipments of LMOs[11]. This principle applies to LMOs for intentional introduction

---

[7] The Doha Work Programme emerged from the Declarations adopted by WTO Members at the 4th WTO Ministerial Conference held in Doha, Qatar on 9-13 November 2001.

[8] See: Oxford Analytica, *Biosafety treaty enters into force*, 29 September 2003.

[9] The use of the term LMO – instead of GMO – was preferred by some delegations that find it more precise and devoid of negative connotations.

[10] The formulation of the precautionary principle contained in Principle 15 of the Rio Declaration of 1992 is the following: "Where there are threats of serious or irreversible damage, lack of full scientific certainty shall not be used as a reason for postponing cost-effective measures to prevent environmental degradation".

[11] One of the central points of contention during the negotiations of the Protocol was whether, in the presence of significant scientific uncertainty, the precautionary approach would represent an appropriate basis on which to take decisions. The Miami Group – which included the main producers and exporters of genetically modified seeds and crops, namely Argentina, Australia, Canada, Chile, the United States and Uruguay - and industry called for all decisions under the Protocol to be based on science, on the assumption that the potential risks posed by LMOs were already well known. According to them, to rely on the precautionary principle would open the Protocol to abuses and trade protectionism and to potential tensions with the SPS Agreement. The EU, the Like-Minded Group – which consisted of the large majority of developing countries - consumer and

into the environment, as well as to those for direct use as food, feed, or for processing. The precautionary approach is one of the main features of the Protocol and reference is made to it in the Preamble, in Article 1 ("Objective"), and in Articles 10 and 11. It allows importing countries to ban imports because of lack of scientific certainty. The ban may last until the importing country decides that it has arrived at scientific certainty about the effects of the products on biodiversity and human health. However, since the importing country is not obliged to seek the information necessary for reaching scientific certainty, a trade-restrictive measure may be in force without time limits. On the contrary, the SPS Agreement allows countries to provisionally adopt sanitary or phytosanitary measures when relevant scientific evidence is insufficient, but obliges them to seek the additional information necessary for a more objective assessment of risk and to review the SPS measure within a reasonable period of time.

For LMOs for intentional introduction into the environment, the Protocol allows the exporting country to request the importing country to review a decision it has taken when a change in circumstances has occurred that may influence the outcome of the risk assessment upon which the decision was based, or additional relevant scientific or technical information has become available. The importing country must respond to such a request in writing within 90 days and set out the reasons for its decision (Article 12, para 2 and 3). This provision therefore gives the exporter the right to request the importer to review its decision in the light of new information; however, the importer retains the flexibility to confirm its previous decision, but it has to justify so doing. This discipline echoes the need for review contained in the SPS Agreement when precautionary measures are used, although there are some basic differences: in the case of the SPS Agreement, the country implementing the measure is obliged to seek additional information and review the SPS measure within a reasonable period of time. In the case of the Protocol, the country implementing a restrictive measure is obliged only to consider the request made by the exporter, analyse the new circumstances or the new scientific or technical information brought to its attention and give a justified reply within 90 days. Moreover, this rule does not apply to LMOs for direct use as food, feed or for processing.

The Protocol states that risk assessment should be carried out in a scientifically sound manner in order to identify and evaluate the possible adverse effects of LMOs on the conservation and sustainable use of biological diversity, taking into accounts risks to human health (Article 15). Article 16 deals with mechanisms, measures and strategies to regulate, manage and control the risks identified by risk assessment.

Article 15 requires the Party of import to ensure that risk assessments are the basis for reaching decisions on proposed imports of LMOs for intentional release into the environment. The Party of import may carry out the risk assessment – often on the basis of the information provided by the potential exporter – or request the exporter to do so. If the risk assessment is performed by the importer, s/he can recover the cost of it from the potential exporter. Risk assessment is also to be used for LMO-FFPs and is among the necessary information to be provided to the Biosafety Clearing-House by a Party that takes a final decision regarding domestic use of LMO-FFPs that may be subject to transboundary movement. Developing Parties or Parties with an economy in transition which have not yet developed a domestic regulatory framework on biosafety, may declare that their decision prior to the first import of LMO-FFPs will be taken according to a risk assessment.

---

green groups, on the other hand, argued that while scientific input remained essential in the field of biosafety, risks posed by LMOs were still not fully understood and could be potentially irreversible. Therefore, the possibility of taking a precautionary approach was seen as crucial for the decision-making regime under the Protocol. The final text of the Protocol includes elements from the different negotiating groups; it goes, however, more in the direction of the EU's and the Like-Minded Group's approach.

In dealing with the same issue – risk assessment and risk management – the SPS Agreement states that sanitary and phytosanitary measures should be based on an assessment of the risks to human, animal or plant life or health. It is interesting to note that under the Agreement a country can base its measures on the risk assessments carried out by other countries or by international organizations, and may seek additional information from other Member countries or from the industry. Members are free to decide the level of risk they are willing to tolerate, however, the Agreement includes the objective of minimizing negative trade effects: measures should not be more trade restrictive than necessary to achieve the chosen level of protection. This means that if there is an alternative measure, equally effective to achieve the appropriate level of protection, that is reasonably available from a technical and economic point of view, that measure should be used. The SPS Agreement embodies also the obligation for Members to avoid arbitrary or unjustifiable distinctions in the levels of sanitary and phytosanitary protection they consider to be appropriate in different situations, if such distinctions result in discrimination or disguised restriction on international trade.

Two aspects of the discipline on risk assessment and risk management, respectively developed within the Biosafety Protocol and the WTO framework, may then be in tension with each other. While, under the Protocol, it is the obligation of the Party of import to ensure that its decision is based on a risk assessment, it may require the exporter to carry out or bear the costs of the risks assessment. In the case of the SPS Agreement, on the other hand, it is the country of import which has to prove, through a risk assessment, the need for the sanitary and phytosanitary measures it has in place. It has to bear the costs of the risk assessment. Secondly, the SPS Agreement includes reference to the restrictive trade impact that a sanitary or phytosanitary measure may have and call for minimizing it. In the Biosafety Protocol this preoccupation is not address.

Under Article 26 of the Biosafety Protocol, Parties may take into account, when deciding whether and under which conditions allow the import of LMOs, "socio-economic considerations arising from the impact of LMOs on the conservation and sustainable use of biological diversity, especially with regard to the value of biological diversity to indigenous and local communities". It then appears that the Protocol would allow trade restrictive measures justified by the fact that imports of LMOs might lead to a loss of cultural traditions, knowledge and practices, particularly among indigenous and local communities. Within the SPS framework, risk assessment can, in specific cases, take into account socio-economic considerations. This happens for the assessment of risks to animal or plant life or health where Members are to take into account relevant economic factors, such as the potential damage in terms of loss of production or sales in the event of the entry, establishment or spread of a pest or disease; the costs of control or eradication in the territory of the importing Member; and the relative cost-effectiveness of alternative approaches to limiting risks (Article 5.3). The same solution does not apply, however, to the assessment of risk to human health. In an early dispute, a GATT Panel rejected trade restrictions that were solely justified on the grounds that cheap imports would undermine the traditional livelihoods of a certain minority population.[12]

3. The WTO Agreements which have direct implications for international trade in GMOs

Four WTO Agreements appear to have special relevance for international trade in GMOs: the SPS Agreement, the TBT Agreement, the TRIPS Agreement, and the GATT 1994. It is not the purpose of this paper to provide a detailed analysis of such Agreements, but only to analyze those provisions that may be in tension with the discipline included in the Biosafety Protocol.

---

[12] *Japanese Measures on Imports of Leather*, GATT Panel Report BISD 31S/94, 2 March 1984, p. 44. On the issue of socio-economic considerations see: *An Explanatory Guide to the Cartagena Protocol on Biosafety, supra*, endnote 4, at 238-239.

Concerning the SPS Agreement, its main goal is to prevent domestic sanitary or phytosanitary measures from having unnecessary negative effects on international trade and being misused for protectionist purposes. However, the Agreement fully recognizes the legitimate interest of countries in setting up rules to protect food safety and animal and plant health, and in fact allows countries to give these objectives priority over trade, provided there is a demonstrable scientific basis for their food safety and health requirements.[13] Although the Agreement does not refer to GMOs explicitly, it can be argued that measures aimed at regulating such a trade could reasonably come within the scope of the Agreement.[14]

As mentioned above, there is limited scope to apply the precautionary principle under the SPS Agreement, contrary to the Biosafety Protocol. The Agreement permits the adoption of SPS measures on a provisional basis as a precautionary step in cases where there is an immediate risk of the spread of pests or diseases but where the scientific evidence is insufficient. However, "Members shall seek to obtain the additional information necessary to a more objective assessment of risk and review the sanitary or phytosanitary measure accordingly within a reasonable period of time" (Article 5.7, second sentence).

In the well-known *hormone* case,[15] related to a ban imposed by the EC on bovine meat and meat products from cattle treated with growth hormones, the role of the precautionary principle in the framework of the SPS Agreement was addressed.

The EC invoked the precautionary principle in support of its claim that its measures were based on a risk assessment. Its basic submission was that the precautionary principle was or had become a "general customary rule of international law" or at least a "general principle of law". Referring more specifically to Article 5.1 and 5.2 of the SPS Agreement, the EC reached the conclusion that since applying the precautionary principle meant that it was not necessary for all scientists around the world to agree on the possibility and magnitude of the risk, or for all or most of the WTO Members to perceive and evaluate the risk in the same way, its measures (an import ban) were precautionary in nature and satisfied the requirements of Article 2.2 and 2.3 of the Agreement, as well as the requirements of paragraphs 1 to 6 of Article 5. According to the United States, on the other hand, the precautionary principle did not represent customary international law: it was more an approach than a principle. For Canada, the precautionary approach or concept was "an emerging principle of law", but had not yet been incorporated into the corpus of public international law. The Panels concluded that the precautionary principle had not been written into the SPS Agreement as a ground for justifying SPS measures that were otherwise inconsistent with the obligations of Members set out in particular provisions of the Agreement and that it did not by itself, and without a clear textual directive to that effect, relieve a panel of the duty of applying the normal (i.e. customary international law) principles of treaty interpretation in reading the provisions of the SPS Agreement. The Appellate Body[16] stated that it was unnecessary, and probably imprudent, for it to take a position on the important but abstract question of the status of the precautionary principle in international law. However, it appeared important to note some aspects of the relationship of the precautionary principle with the SPS Agreement. The Appellate Body upheld the Panels' conclusions

---

[13] More specifically, the Agreement covers measures adopted by countries to protect human or animal life from food-borne risks; human health from animal- or plant-carried diseases; animal and plants from pests and diseases; and the territory of a country from the entry, establishment, or spread of pests.

[14] As of 30 July 2003, 82 notifications related to biotech products had been submitted to the WTO secretariat under the notification system established by the SPS Agreement.

[15] *EC Measures Concerning Meat and Meat Products (Hormones)*, Complaint by the United States, WT/DS26/R, 18 August 1997; Complaint by Canada, WT/DS48/R, 18 August 1997.

[16] WT/DS26/AB/R, WT/DS48/AB/R, 16 January 1998.

that the precautionary principle would not override the explicit wording of Article 5.1 and 5.2 and stressed that it had been incorporated into Article 5.7 of the SPS Agreement, but this provision did not exhaust the relevance of the precautionary principle for SPS.

In the *hormone* case the Panels and the Appellate Body did not have a chance to interpret directly Article 5.7 of the SPS Agreement, because the EC had not invoked it to justify the measures in dispute. However, Article 5.7 of that Agreement was explicitly addressed in the *Japan varietals* case.[17] The case was about a complaint by the United States relating to the requirement imposed by Japan for testing and confirming the efficacy of the quarantine treatment for each variety of certain agricultural products. In support of its varietal testing requirement, Japan invoked Article 5.7. According to the Appellate Body, Article 5.7 sets out four cumulative requirements that must be met for adopting and maintaining provisional SPS measures. A country may provisionally adopt an SPS measure if this measure is: (i) imposed in respect of a situation where relevant scientific information is insufficient; and (ii) adopted on the basis of available pertinent information. Such a measure may not be maintained unless the country that adopted it: (i) seeks to obtain the additional information necessary for a more objective assessment of risk; and (ii) reviews the measure accordingly within a reasonable period of time.

It seems, therefore, that the WTO jurisprudence is proposing a rather narrow interpretation of Article 5.7 of the SPS Agreement: by stressing the need for countries to comply with four specific requirements in order to be able to invoke the right to adopt and maintain provisional measures, and by stating that the precautionary principle would not override the need for countries to base their SPS measures on a risk assessment - and, in general, by avoiding the expression of any view on the status of the precautionary principle in public international law. However, the Appellate Body also stated that Article 5.7 did not exhaust the precautionary principle for SPS. It seems that the central role of scientific evidence and risk assessment as the necessary bases for taking and maintaining SPS measures is reconfirmed. While the precautionary principle may be invoked to justify time-limited measures, it does not represent a long-term alternative to risk assessment and scientific evidence.

The analysis will now turn to consider potential overlaps between the TBT Agreement and the Biosafety Protocol. It is noteworthy that the TBT and the SPS Agreements, as well as the GATT 1994, can apply to international trade in GMOs. So far as the relationship among them is concerned, once SPS applies, TBT cannot apply. In the event of conflict between SPS/TBT and GATT, the specific agreement prevails over GATT, according to the General Interpretative Note to Annex 1A.

Labelling requirements related to food, nutrition claims and concerns, quality and packaging regulations are normally subject to the TBT Agreement. While SPS measures may be imposed only to the extent necessary to protect human, animal or plant health from food-borne risks or from pests or diseases, Governments may introduce TBT regulations when necessary to meet a number of legitimate objectives, including the prevention of deceptive practices, the protection of human health or safety, animal or plant life or health, or the environment. Technical regulations should not create unnecessary obstacles to international trade and be more trade-restrictive than is necessary to fulfil a legitimate objective, taking account of the risks that non-fulfilment would create. Measure should not discriminate between imported products and "like" products of domestic or foreign origin. If GMOs and GM products are considered "like" products in relation to conventional products, there are no grounds for applying any special treatment to them, including mandatory labelling schemes.

---

[17] *Japan - Measures Affecting Agricultural Products*, WT/DS76/R, 27 October 1998, and WT/DS76/AB/R, 22 February 1999.

Turning to the GATT 1994, the national treatment principle, which is incorporated into Article III, implies non-discrimination between domestic and imported goods. Translating this principle into the GMO context implies that the importing country is not allowed to apply to foreign products measures more onerous than those applied to like domestic products. In the context of Article III as well, the determination of what constitutes "like products" is a crucial issue since the national treatment obligations apply only if two products are "like". In assessing whether products are "like", the controversial issue of whether the analysis should be limited to the physical characteristics of the products or should also take into account the process and production methods is still open. The relevant jurisprudence is not conclusive and authoritative authors are deeply divided on that subject.[18] On the one hand, it has been argued that there is no real support in the text and jurisprudence of the GATT for the product/process distinction,[19] and that the distinction is neither warranted nor useful in practice.[20] On the other hand it has been suggested that there is a textual basis in GATT Article III and the Note ad Article III for the product/process distinction and that the distinction should be retained to prevent protectionist abuses.[21] The product/process distinction is therefore an open issue.

The general elimination of quantitative restrictions is embodied in Article XI of the GATT 1994, which provides that no prohibitions or restrictions other than duties, taxes or charges shall be instituted or maintained on the importation or exportation of any product.

The obligations of Articles III and XI can be derogated from by using the exceptions set out in Article XX of the GATT 1994. The provisions of the latter which are of special relevance for trade in GMOs are as follows:

General Exceptions

Subject to the requirement that such measures are not applied in a manner which would constitute a means of arbitrarily or unjustifiable discrimination between countries where the same conditions prevail, or a disguised restriction on international trade, nothing in this Agreement shall be construed to prevent the adoption or enforcement by any contracting party of measures:

....

(b) necessary to protect human, animal or plant life or health;

.....

(g) relating to the conservation of exhaustible natural resources if such measures are made effective in conjunction with restrictions on domestic production or consumption.

....

Article XX gives countries the legal means to balance their trade obligations with important non-trade objectives, such as health protection or the preservation of the environment, which form part of their

---

[18] However, it has been stressed that the "trade policy elite has simply accepted the notion of a sharp divergence between measures on products and PPMs as if such a distinction had been written into the GATT all along, and not simply invented in the *Tuna/Dolphin* case": Trebilcock M.J. and R. Howse, *The Regulation of International Trade* (London and New York: Routledge, 1999) at. 413.

[19] Howse, R. and D. Regan, "The Product/Process Distinction - An Illusionary Basis for Disciplining 'Unilateralism' in Trade Policy", *European Journal of International Law* 11, 2000, pp. 249 ff., at 264-268.

[20] Cosbey, A., "The WTO and PPMs: Time to Drop a Taboo", *Bridges* 5 No. 1-3, 2001, pp.11-12.

[21] Jackson, J.H., "Comments on Shrimp/Turtle and the Product/Process Distinction", *European Journal of International Law* 11, 2000, pp. 303-307.

overall national policies. In the *Shrimp* case[22] the Appellate Body, referring to the introductory text of Article XX, stated that "[W]e consider that it embodies the recognition on the part of WTO Members of the need to maintain a balance of rights and obligations between the right of a Member to invoke one or another of the exceptions of Article XX, specified in paragraphs (a) to (j), on the one hand, and the substantive rights of the other Members under the GATT 1994, on the other hand... A balance must be struck between the right of a Member to invoke an exception under Article XX and the duty of that same Member to respect the treaty rights of the other Members".[23] According to the Appellate Body, the purpose of the introductory text of Article XX is "generally the prevention of the abuse of the exceptions of Article XX".[24]

Turning to the last WTO Agreement that has a direct relevance for international trade in GMOs, namely the TRIPS Agreement, analysis will be again limited only to the potential tensions between the discipline included in the Agreement and that of the Biosafety Protocol.

Strengthened protection of intellectual property rights may make investment by the biotechnology industry more profitable.[25] The TRIPS Agreement, then, may be seen as promoting the adoption of GMOs in the food system. Related to the issue of biotechnology applied to agricultural and food products is the issue of obtaining patents on live plants or animals, including biotechnological inventions and plant varieties. Concerns are being expressed in both developed and developing countries about the economic, social, environmental and ethical impacts of life patenting. Moreover, many developing country Governments are concerned that the control of the nature and distribution of new life forms by transnational corporations may affect their countries' development prospects and food security.

Currently, the TRIPS Agreement does not require that countries grant patents for plants and animals; however, they have to provide for the protection of plant varieties either by patents or by an effective *sui generis* system,[26] or by a mixture of both (Article 27.3(b)). The revision of Article 27.3(b) is part of the "built-in agenda" agreed at the conclusion of the Uruguay Round. In accordance with that agenda, the WTO Council for TRIPS started the revision of Article 27.3(b) in 1999; however, owing to lack of consensus among Members, the revision is still ongoing.[27]

Considering that GMOs and GM crops incorporate patented technology and that trade-restrictive measures implemented under the Biosafety Protocol affect those products, it may be argued that those

---

[22] Appellate Body Report on *United States-Import Prohibition of Certain Shrimp and Shrimp Products*, adopted on 12 October 1998, WT/DS58/AB/R.

[23] *Shrimp*, at para. 156.

[24] *Shrimp*, at para. 150.

[25] For a detailed analysis of the issue see: Tansey G., "Trade, intellectual property, food and biodiversity", Discussion Paper, Quaker Peace & Service, London, February 1999.

[26] A *sui generis* system of protection is an alternative, unique form of intellectual property protection, designed to fit a country's particular context and needs. In the case of plant varieties, it means that countries can make their own rules to protect new plant varieties with some form of intellectual property rights (IPRs), provided that such protection is effective. The Agreement does not define the elements of an effective system.

[27] Some developing countries have proposed to amend the TRIPS Agreement to require patent applicants to disclose the source of origin of the biological resources, and to provide evidence of prior informed consent and benefit sharing. The African Group has called for Article 27.3(b) to be revised so as to prohibit patenting of plants, animals and micro-organisms. Switzerland would like to see these issues discussed outside the WTO and moved to the World Intellectual Property Organization (WIPO). The EC would support mandatory disclosure of origin requirements; such a requirement, however, should not constitute an additional patentability criterion and failure to disclose should be regulated by civil or administrative law. See: ICTSD and IISD, Doha Round Briefing Series, *Intellectual Property Rights*, Vol.2 No.5, August 2003.

measures may nullify or impair Members' rights under TRIPS.[28] An additional concern is the degree to which patent holders and licensees will be responsible and liable for any adverse consequences of the application of biotechnology for the environment and human well-being.

4. Actual and potential GM-related trade disputes

The Preamble of the Biosafety Protocol states that it shall not be interpreted as implying a change in the rights and obligations of the parties under existing international agreements and that this recital is not intended to subordinate the Protocol to other international agreements. These provisions may prove not to be very helpful if a conflict arises between countries with divergent interests in the area of biotechnology. Disputes may occur between parties to the Protocol, for instance on the interpretation of the role that the precautionary approach can play in decision-making, or between parties and non-parties on such issues as import restrictions, notification and identification requirements, delays in evaluating requests and authorize imports, or on special conditions attached to the imports, such as mandatory labelling requirements.

Countries which are parties to a multilateral agreement are supposed to solve their possible conflicts within the framework of the agreement they have signed and ratified. However, in the case of the Biosafety Protocol, if a party believes that in a specific circumstance its interests are better protected by WTO rules[29], it may invoke those rules, arguing that the Protocol clearly states that it shall not be interpreted as implying a change in the rights and obligations of the parties under existing international agreements. A possible conflict *between parties* may therefore be settled under the WTO dispute settlement mechanism. It flows from Article 23 of the Dispute Settlement Understanding that any WTO Member can initiate a case in the WTO if it considers that its market access rights have been violated.

On the other hand, a country which has an interest in solving a dispute according to the discipline laid down in the Biosafety Protocol may invoke the fact that the Protocol represents *lex specialis,* which has priority over *lex generalis* (WTO agreements). It may also refer to the principle that later in time prevails. Finally, it could ask for its WTO obligations to be interpreted in the light of the Protocol. The WTO Appellate Body stated in two disputes[30] that the WTO legal system does not operate in "clinical isolation" from existing rules of public international law. Therefore, non-WTO treaties, practices, customs and general principles of law may be relevant in the interpretation of WTO provisions and can become quite influential in defining the parameters and the content of WTO obligations. Then, the Protocol will likely have a role to play.

---

[28] A "non violation nullification or impairment" measure is one which, while it does not conflict with the provisions of the Agreement, has the effect of nullifying or impairing a "benefit" ensured under a treaty. The rationale of such a provision is to protect the overall balance of concessions reasonably expected when the agreement was reached. This concept finds its roots in trade in goods. Article 64.3 of the TRIPS Agreement, by referring to Article XXIII of the GATT, opens the possibility of applying non-violation complaints in the field of IPRs. However, the extension of the non-violation discipline to IPRs should be agreed by WTO Members by consensus. Consensus is still missing.

[29] Assuming that all countries involved are Members of the WTO.

[30] In the *Gasoline* case (*United States – Standards for Reformulated and Conventional Gasoline*, WT/DS2/R, 29 January 1996), the Appellate Body cited Article 3.2 of the Dispute Settlement Understanding, which requires Panels and the Appellate Body to use "customary rules of interpretation" to interpret the provisions of the WTO Agreements. The Appellate Body linked the WTO legal system to the rest of international legal order and imposed on Panels and WTO Members the obligation to interpret the WTO Agreements in accordance with customary rules of interpretation of public international law. In the *Shrimp* case (*supra*, footnote 21), the Appellate Body made reference to various international conventions to interpret the term "natural resources". For a detailed discussion on this topic see Marceau G., "A Call for Coherence in International Law: Praises for the Prohibition Against 'Clinical Isolation' in WTO Dispute Settlement", *Journal of World Trade*, October 1999, pp. 87 ff.. See also: J. Pauwelyn, "The Role of Public International Law in the WTO: How Far Can We Go?", *American Journal of International Law*, Vol. 95, No.3, 2001, pp. 535 ff..

If a dispute occurs between *a party and a non-party* to the Protocol, the case will most likely be brought to the attention of the WTO Dispute Settlement Body.

The issue of the relationship between the trade rules included in multilateral environmental agreements (MEAs) and WTO rights and obligations, and in particular the issue of which rules would prevail in case the trade provisions of a MEA conflict with WTO rules has been discussed for several years in various international forums, without any conclusive result. A related unsolved issue is the position on non-parties to a multilateral agreement which may be affected by the trade rules agreed by the parties to a multilateral agreement. The Doha Ministerial Declaration mandated, in para 31(i), negotiations on the relationship between existing WTO rules and specific trade obligation set out in MEAs. WTO Members have agreed in the WTO Committee on Trade and Environment to analyze the "specific trade obligations" in six MEAs, including the Biosafety Protocol, with a view of examining their interaction with relevant WTO rules. Beyond this factual decision, however, no substantive progress on the issue was made. The topic received almost no attention at the Cancun Ministerial Conference and the draft ministerial declaration (not adopted at the end of the meeting) contained a statement only aimed to "reaffirm our commitment to these negotiations".[31]

Even though the trade provisions of a multilateral agreement have not yet been challenged before a WTO dispute settlement panel, it may be argued that the Biosafety Protocol is different from other multilateral agreements and that there is a more concrete risk that its WTO compatibility may be challenged. This is because the economic interests involved in international trade in GMOs are huge; public opinion is still very much divided on whether biotechnology is a risk or an opportunity; the United States, which is the single largest producer of GM crops, on one hand has actively participated in the negotiations, but, on the other hand, is not a party to the Protocol and very unlikely will join it (since it has not ratified the CBD), and the Protocol is already being interpreted in divergent ways.

A few weeks before the Biosafety Protocol entered into force, three complains about EC restrictive measures affecting GMOs and GM crops were officially brought to the attention of the WTO Dispute Settlement Body. More or less at the same time the EC enacted new legislation on GMOs and GM food.[32]

On 7 August 2003, the United States, Canada and Argentina each requested the establishment of a panel on "Measures affecting the approval and marketing of biotech products".[33] According to them, since October 1998 the EC had applied a de facto moratorium on the approval of products of agricultural biotechnology. This moratorium had led to (i) the suspension of and failure to consider various applications for approval of agricultural biotechnology products; and (ii) undue delays in finalizing the processing of applications for the approval of such products. The complaining countries also alleged that several EC member States[34] had introduced bans on the importation, marketing or sale of a number of biotech products which had already been approved at Community level, thereby infringing both WTO rules and Community legislation. The approvals moratorium and the national marketing and import bans had allegedly restricted imports into the EC of agricultural and food products from the United States, Canada and Argentina.

---

[31] Draft Cancun Ministerial Text, Second Revision, 13 September 2003.

[32] On this issue see: D. Brack, R. Falkner and J. Goll, *The next trade war? GM products, the Cartagena Protocol and the WTO*, The Royal Institute of International Affairs, Sustainable Development Programme, Briefing Paper No. 8, September 2003.

[33] WTO documents: WT/DS291/23; WT/DS292/17; WT/DS293/17 dated August 2003.

[34] Austria, France, Germany, Greece, Luxembourg and United Kingdom.

At the beginning of the 1990s, the EC introduced an approval system for the deliberate release into the environment of GMOs for experimental purposes or as commercial products, with the aim of ensuring a high and uniform level of protection of health and the environment throughout the Community and the efficient functioning of the internal market.[35] The 1990 Directive was later replaced by Directive 2001/18, which entered into force on 17 October 2002.[36]The new directive establishes harmonized procedures and criteria for the case-by-case evaluation of potential risks arising from the deliberate release of GMOs into the environment. All GM seed varieties have to be approved and authorized in the EU for cultivation. The Directive requires that applications must be accompanied by: a full environmental risk assessment, detailed information on the GMO, its release conditions, interaction with the environment, monitoring, waste and contingency plans, labelling and packaging proposals. The Directive further provides for a rather complex approval procedure involving both national competent authorities and Community bodies, including a brand new European Food Safety Authority (EFSA).

In addition to "horizontal" legislation, the EC has adopted a number of "vertical" Directives and Regulations. This "vertical" legislation is product-oriented, and deals with specific aspects or products resulting from genetic modification. Legislation related to novel foods and novel food ingredients is part of the "vertical" regulatory approach.[37] It stipulates that, in order to protect public health, guarantee the proper functioning of the internal market and create conditions of fair competition, it is necessary to ensure that novel foods and novel food ingredients[38] are subject to a single safety assessment through a Community procedure before they are placed on the EC market. The regulation establishes a system of authorization for placing on the market novel foods, including food products containing, consisting of or produced from GMOs.

Within the described legal framework, 18 GMOs were authorized for commercial release into the environment and a limited number of GM food products were approved for marketing. However, rising concerns about the possible effects of biotech products on health and on the environment and mounting resistance by European consumers to consume biotech products led to the suspension of all new authorizations pending the adoption of new rules to ensure labelling and traceability. As a result, no new GMOs have been approved in the EC since October 1998. 33 applications for release into the environment (13 submitted under Directive 90/220 and, after its repeal, 20 submitted under Directive 2001/18) and 10 applications for food products were frozen[39]. The moratorium has had an obvious impact on the exports of biotech products and the United States, being the largest producer of GMOs and the owner of most of the related technology, has been particularly affected.

On 22 July 2003, Community institutions agreed on two new regulations which establish a system to trace and label GMOs and to regulate the placing on the market and labelling of food and feed products derived from GMOs. This new legislation will ensure full traceability of GMOs (i.e. the tracking of the

---

[35] Council Directive 90/220/EEC, 23 April 1990, OJ L 117, 8 March 1990, pp. 15 ff.

[36] Directive 2001/18/EC of the European Parliament and of the Council of 12 March 2001 on the deliberate release into the environment of genetically modified organisms and repealing Council Directive 90/220/EEC, OJ L 106, 17.04.2001. Directive 2001/18 was amended by Regulation 1830/2003. See the EC website on GMOs at: http://www.europa.eu.int/comm/food/fs/gmo/gmo_index_en.html.

[37] Regulation (EC) No. 258/97, 27 January 1997, OJ L 043, 14 February 1997, pp. 1 ff.

[38] Under the Regulation, novel foods and food ingredients are those which have not yet been used for human consumption to a significant degree within the Community, in particular those containing or derived from GMOs.

[39] Memo/020160 – Rev., Brussels, 1 July 2003. Found at: www.europa.eu.int/comm/dgs/health_consumer/library/press/press298_en.pdf

movement of GM products) throughout the production and distribution chains and will provide consumers with comprehensive information by labelling all food and feed consisting of, containing or produced from a GMO.[40] These two regulations complete the previously enacted legislation in this area and, as of July 2003, the EU legislation on GMOs may be said to be complete. The two new regulations address the most pressing concerns of the public regarding the environmental and health effects of GMOs, providing a reinforced safeguard system and enabling consumers to choose, through comprehensive, compulsory labelling. Once the regulations enter into force, it is expected that the de facto moratorium on GMO approval will be lifted. Nevertheless, it is reported that some EC Member governments have hinted that they will block a restart of the approval process until strict environmental liability legislation is also in place. [41]

The new rules implemented by the EC have not, however, appeased US, Canadian and Argentinean farmers and industry trade bodies, which, on the contrary, have pushed the respective administrations to take further steps in the context of the WTO dispute settlement procedures.

Further to the complains presented by the United States, Canada and Argentina, the WTO Dispute Settlement Body established a single panel at its meeting on 29 August 2003. Argentina, Australia, Brazil, Canada, Chile, China, Colombia, El Salvador, Honduras, Mexico, New Zealand, Norway, Peru, Chinese Taipei, Thailand, Uruguay and the United States reserved their third-party rights.

The complaining countries alleged that the measures at issue appeared to be inconsistent with the EC's obligations under the SPS and TBT Agreements, the Agreement on Agriculture and the GATT 1994. More specifically, they alleged violation (some paragraphs or the entirety) of Articles 2, 5, 7 and 8, and Annexes B and C of the SPS Agreement[42]; Articles I, III, X and XI of the GATT 1994[43]; Article 4 of the Agriculture Agreement[44]; and Articles 2 and 5 of the TBT Agreement[45]. Argentina additionally alleged violation of Article 10 of the SPS Agreement and Article 12 of the TBT Agreement, while Canada also alleged violation of Article XXIII:1(b) of the GATT[46].

---

[40] Regulation (EC) No 1829/2003 of the European Parliament and of the Council of 22 September 2003 on genetically modified food and feed; OJ L 268, 18/10/2003, pp. 1-23; and Regulation (EC) No 1830/2003 of the European Parliament and of the Council of 22 September 2003 concerning the traceability and labelling of genetically modified organisms and the traceability of food and feed products produced from genetically modified organisms and amending Directive 2001/18/EC; OJ L 268, 18/10/2003, pp. 24-28. See also European Commission Press Release IP/03/1056, *European legislative framework for GMOs is now in place*, 22 July 2003.

[41] For a more detailed description of the relevant EC legislation, see: UNCTAD, *Training Module on the WTO Agreement on Sanitary and Phytosanitary Measures*, Commercial Diplomacy Programme, forthcoming.

[42] Article 2 of the SPS Agreement states the basic rights and obligations under the Agreement. Article 5 deals with the assessment of risks and the appropriate level of sanitary or phytosanitary protection and includes the possibility for Members to provisionally adopt SPS measures. Article 7 and Annex B set out the obligations related to transparency, while Article 8 and Annex C deal with control, inspection and approval procedures.

[43] Article I and III include the cornerstone principles of the Most-Favoured Nations (MFN) and of National Treatment (NT). Article X refers to the transparency obligation for Members to publish promptly trade laws and regulations affecting international trade. Article XI refers to the obligation of general elimination of quantitative restrictions.

[44] Article 4 states that Members shall not maintain any of those border measures - such as quantitative import restrictions, import licensing, minimum import prices - which have been required to be converted into customs duties.

[45] Article 2 sets out the rules related to the preparation, adoption and application of technical regulations and includes the concept of "like products". Article 5 spells out the obligations regarding conformity assessment procedures, including that they have to be undertaken and completed as expeditiously as possible.

[46] Article 10 of the SPS Agreements and Article 12 of the TBT Agreement on Special and Differential Treatment refer to the obligation to take into account the special needs of developing countries in the preparation and application of SPS/TBT measures. Article XXIII:1(b) of the GATT 1994 refers to non-violation complains.

A country that is banning or otherwise putting under restrictions imports of GMOs or GM products may be infringing its trade obligations under the WTO; it can, however, invoke a number of provisions to justify its trade-restrictive measure. It may invoke the SPS Agreement. This is because measures related to GMOs may have the goal of protecting "human or animal life from risks arising from additives, contaminants, toxins or disease-causing organisms in their food, beverages, feedstuffs", and/or "a country from damage caused by the entry, establishment or spread of pests" (in view of the lack of scientific certainty about the impact of GMOs on the environment, avoiding the transfer of genetic material and associated traits from engineered varieties to conventional varieties could be regarded as similar to protecting plants from pests and diseases). In other words, measures related to GMOs may fall within the spirit, if not the letter, of the SPS Agreement. The country imposing the trade restrictive measure has to prove that it is necessary to protect human, animal or plant life or health, it is based on scientific principles and it is not maintained without sufficient scientific evidence. If the measure is applied on a provisional basis, it must seek to obtain the additional information necessary for a more objective assessment of risk and review the measure accordingly within a reasonable period of time. There may be some difficulties at present in invoking the SPS Agreement to justify a trade-restrictive measure in respect of GMOs. There is no scientific evidence that clearly identifies the risk that GMOs create for human, animal or plant life or health. If the measure at stake is a trade ban, alternative measures, less trade restrictive, may be available. If a measure is taken on the basis of the precautionary principle, it has to be reviewed within a reasonable time frame. However, if the justification of the trade-restrictive measure is not safety, the SPS Agreement is not applicable and not violated.

A second option to justify a GM trade-restrictive measure is to invoke the TBT Agreement. Based on the stated objectives, an import ban applying to GMOs or GM products could likely be regarded as a technical regulation falling under the TBT Agreement. According to the jurisprudence in the *Asbestos* case[47], a measure which lays down product characteristics – for instance that the food does not contain any material or ingredient that has been subjected to genetic manipulation, including the applicable administrative provisions – is a technical regulation. The requirement that measures has to be not more trade-restrictive than necessary and the linked "proportionality test" (i.e. the need to establish a kind of balance between the restrictive trade impact of a measure and the risks that non-fulfilment of the stated objectives would create), seem to be relevant in the framework of international trade in biotech products. At the same time, if the stated objective of a measure is the protection of human health or safety, animal or plant life or health, or the environment, the application of the proportionality test would seem to be particularly problematic, considering that there are at present very divergent views on the magnitude of the risk that GMOs might create. On the other hand, some argue that there is no proportionality test included in the TBT Agreement and the issue is only whether the measure chosen is not unnecessarily trade-restrictive, considering the level of protection that a country has chosen. In that case, a country could implement strict technical regulations regarding GMOs, even though the regulations might have a considerable trade-restrictive impact, provided that they are not more trade-restrictive than necessary.

Another relevant aspect of the TBT Agreement is the concept of "like products". Article 2.1 of the TBT restates the principle of non-discrimination set forth in Article I:1 and Article III:4 of the GATT 1994. If the claimant contends that a technical regulation is incompatible with Article 2.1 of the TBT because it subjects imported genetically modified products to less favourable treatment than conventional products of national or foreign origin, the Panel, in order to determine incompatibility with Article 2.1 of the TBT, would have to establish, *inter alia*, that the genetically modified and conventional products involved are "like products". In this context, it seems that the issue to consider is whether a genetically engineered

---

[47] *European Communities – Measures Affecting Asbestos and Asbestos-containing Products*, Report of the Appellate Body WT/DS135/AB/R, 12 March 2001.

product that sufficiently resembles a conventional product in outward characteristics would be considered substantially equivalent to the conventional product. If this was the case, the two products would therefore be regarded as equally safe and should be treated in the same way.

The issue of "like products" within the framework of international trade in GMOs has already been brought to the attention of the WTO TBT Committee, but it remains open.

Likewise, the issue of labelling of GMOs and GM crops remains open. Some WTO Members consider that informing consumers through labelling of GM products is a legitimate objective that justifies a trade restriction within the TBT discipline. Others argue that, given the emotional character of the debate, labelling would stigmatise GM products, mislead consumers to think that GM-products may be unsafe or substantially different from conventional counterparts. Because the legitimacy of mandatory labelling systems relates to the definition of "like" products, it is unlikely to be solved by the TBT Committee. Provisions on information to be included in the accompanying documentation of GMOs and genetically modified commodities have been included in the Biosafety Protocol and will be developed further, however, the problem of the consistency of these provisions with those of the TBT Agreement has not been addressed.

A third option to justify trade restrictive measures affecting biotechnology products is to invoke the GATT. In this case the first question to clarify is whether the measure at stake is an import ban (Article XI of GATT) or an internal regulation enforced at the point and time of importation (Article III:4 and Note Ad Article III of GATT). In most cases, in fact, domestic GMO regulatory schemes consist of internal regulations enforced, in the case of imported products, at the point of importation through quantitative restrictions.

The categorization of internal measures, which are externally enforced, as "internal regulations" (Article III) or "quantitative restrictions" on importation (Article XI), has important regulatory implications. While restrictions on importation are prohibited by Article XI:1, Members are permitted by Article III:4 to impose an internal regulation on products imported from other Members provided that it does not discriminate between "like" products. The issue of "likeness" is not at stake in an inquiry under article XI, while it stands at the heart of the analysis under Article III:4.

If the GM-related trade measure is categorised as an internal regulation, and reviewed under Article III:4 of the GATT, it is legitimate under that provision, unless it accords to the "like" imported products "less favourable treatment" than it accords to the "like" domestic products. In this case the crucial question, very similar to that under the TBT Agreement, is whether imported genetically modified organisms and products thereof are "like" their domestic conventional counterparts. According to established GATT practice, the four general criteria which provide a framework for analysing the "likeness" of particular products are: i. physical properties; ii. end-uses; iii. consumers' tastes and habits; iv. and tariff classification.

In the case of GMOs and GM commodities and their conventional substitutes, two of the traditional criteria (i.e., end-use and tariff classification) would point to "likeness", and two (i.e., consumers' perceptions and properties of the products) would point to "non-likeness". The *Asbestos* case may provide some overall guidance when addressing this controversial issue. According to the *Asbestos* jurisprudence, in fact, what is of paramount importance to assess "likeness" under Article III:4 of GATT is the competitive relationship between products in the market place. In this case it may be argued that consumers sense GM-crops and seeds and their conventional counterparts in a different way (because of the perceived negative health and environmental impacts), and that consumers' perceptions and behaviours in respects of the two sets of items ultimately affect the degree of competitiveness or substitutability of GM-crops and seeds and their conventional counterparts in the marketplace. In light of these considerations, it may be difficult to establish that GMOs and their conventional substitutes are "like products".

In the case of processed foods derived from GM-materials and their conventional counterparts, "likeness" seems to be confirmed by the application of at least three of the traditional criteria currently applied to determine whether products are "like", namely, product characteristics, end-use, and tariff classification. In particular, there is a strong physical similarity between processed foods containing GMOs and GMO-free food products, to the extent that the altered molecular or cellular characteristics of the genetically modified organism contained in the product is usually not detectable any longer in the ultimate product. It follows that evidence relating to properties, end-uses, and tariff classification indicates that foods containing GM-materials and their conventional counterparts are "like".

Assuming that, after careful scrutiny of the factual and legal context in the given dispute, the individual GM-products and conventional products are found to be "like" products, there is a second element that must be established before a measure can be held to be inconsistent with Article III:4: "like" imported products are accorded less favourable treatment than "like" domestic products. Only if a "less favourable treatment" is detected - meaning a certain asymmetry between the group of imports as opposed to the group of domestics - the restrictive trade measure can be considered to be in violation of Article III:4. "The term 'less favourable treatment' expresses the general principle, in Article III:1, that internal regulations 'should not be applied…so as to afford protection to domestic production'".[48]

If the measure at stake is found to violate Article XI or Article III:4, it requires justification under Article XX(b) and (g) and the chapeau of Article XX. It is difficult to assess whether or not the measure would be found to come within the scope of Article XX(b) and (g) exceptions. Paragraph (b) requires, *inter alia*, that the measure be "necessary" to protect human, animal or plant life/health. While GATT earlier jurisprudence has interpreted "necessary" as implying a "least-trade-restrictive test", subsequent jurisprudence has not explicitly endorsed it.[49] It appears very difficult, then, to assess how a panel or the Appellate Body would interpret the term "necessary" in a dispute related to a restrictive trade measure affecting GMOs and GM-products. Paragraph (g) requires that the measure be aimed at the conservation of exhaustible natural resources. The panel or the Appellate Body would have then to assess whether the release of GMOs into the environment constitutes a challenge to the conservation of exhaustible natural resources, in particular biological diversity. The way in which the panel and Appellate Body will rule will likely depend very much on the specific facts of each case.

5. Conclusions

Lacking conclusive scientific evidence on the actual or potential impact of agricultural biotechnology on human and animal health and on the environment, the debate continues to be intensive and emotional and countries keep holding rather divergent views about the risks and opportunities that biotechnology may bring about. Those views are reflected in domestic regulations on the approval, marketing and documentation requirements of GMOs and GM products that vary substantially from a country to

---

[48] Ibid., at 100.

[49] In the *Korean – Beef* case (*Korea - Measures Affecting Imports of Fresh, Chilled and Frozen Beef*, WT/DS161, 169/AB/R, 11 December 2000) the Appellate Body held that "…determination of whether a measure, which is not 'indispensable' may nevertheless be 'necessary' within the contemplation of Article XX(d), involves in every case a process of weighing and balancing a series of factors which prominently include the contribution made by the compliance measure to the enforcement of the law or regulation at issue, the importance of the common interests or values protected by that law or regulation, and the accompanying impact of the law or regulation on imports or exports".

another. Domestic regulations have as well an impact on international trade and, then, have to be scrutinized in light of the multilaterally agreed trade rules. The two main legal frameworks applying to the sector are the WTO framework - which is not specific to biotechnology and was actually developed at a time when biotechnology was not an issue - and the Biosafety Protocol which, on the contrary, is a more recent multilateral instrument specifically targeted to GMOs and GM commodities. The two legal frameworks do not seem to be fully consistent with each other. The inability of the international community to rule on how to deal with sectors which are covered by specific multilaterally agreed legal instruments but at the same time are covered by the WTO discipline is *de facto* shifting the responsibility to find a solution from the decision-making level to the dispute settlement level, then, from governments to lawyers/specialists.

These factors - along with the huge economic interests involved in biotechnology and the links that the sector has with biodiversity protection, food security, ethical and religious concerns, human and animal life and health - make the whole issue quite prone to disputes. One was indeed recently brought to the attention of the WTO dispute settlement body.

In the case of trade disputes, it is rather uncertain which legal arguments may prevail. They will likely be different depending on whether GMOs, genetically modified crops or consumer products derived from GM material are at stake. The relevant WTO provisions may be interpreted in a way supporting the reasons of the claimant, as well as those of the defendant. The Biosafety Protocol – which just entered into force - may play a role, but only within its scope (i.e. GMOs and GM commodities) and only with reference to its Parties. It may also play a subsidiary role within the WTO legal system. However, it will be up to the WTO panels and, possibly, the Appellate Body to decide how much legal weight they wish to give to its provisions.

SESSION 3:
RISK ANALYSIS IN DIFFERENT COUNTRIES

SESSION 3.1:
THE EUROPEAN UNION DIMENSION

THE EU LEGISLATION REGARDING GMOs AND ITS IMPLICATIONS FOR TRADE

Denise Prévost*(1)* with **Geert van Calster** *(2)*
(1)Faculty of Law, University of Maastricht
Bouillonstraat 3. Maastricht. The Netherlands.
(2)Faculty of Law, Catholic University Leuven (K.U.L.).
Tiensestraat 41. 3000 Leuven. Belgium.

1. Introduction

The drafting of the new regulations on genetically modified organisms (GMOs) in the European Union (EU) has been a difficult and controversial task. EU regulators have been faced with the challenge of creating rules that address the possible risks to health and the environment from GMOs, in the face of scientific uncertainty regarding these risks. They have also had to ensure that the new rules will give consumers confidence in the safety of those GMOs that are approved for use in the EU. This is part of a general effort to address the erosion of consumer confidence in the EU regulatory system following several recent food-safety crises. As the European Commissioner for Health and Consumer Protection, David Byrne, said in 2001, "Unless we can give EU consumers confidence in this new technology then GM is dead in Europe."

In addition, the new rules on GMOs have to comply with the EU's international trade obligations. In fact, the EU is currently facing a challenge before the World Trade Organisation (WTO) with respect to its (previous) GMO regime.[1] With all these considerations in mind, the EU has revised its legislative regime with regard to the approval and marketing of GMOs. The new legislation was adopted on 22 September 2003 and came into force 20 days after publication thereof in the Official Journal of the EU, on 18 October 2003. Operators will have to comply with the new rules within six months after publication.

This short paper aims to provide a broad overview of the EU legislation applicable to GMOs. It will start by briefly examining the previous EU legislation. It will then explain what is meant by the "de facto moratorium" on approval of GMOs in the EU, before proceeding to a discussion of the new EU legislation. This paper will conclude with a brief overview of the implications of the new EU legislation for business operators wishing to export GM products into the EU.

2. Previous EU Legislation

European Community legislation on GMOs was enacted in the early 1990, and has been progressively refined over the years. This legislation regulated three main issues: authorisation for placing GMOs on the market, labelling of products containing GMOs and traceability of these products. These rules are important for EU traders but also for traders in other countries that want to export GMOs into the EU, as their products will have to meet the requirements set out therein.

---

[1] A WTO panel was established on 29 August 2003 to hear the complaints of the US, Argentina and Canada regarding the EC's suspension of consideration and failure to approve applications for authorisation of GM products and the bans of EC Member States on GM products under the safeguards provisions. It is important to note that the challenge did not focus on the previous EC legislation on GMO approval, but rather the EC inaction in GMO approvals and the Member States' safeguard measures.

Authorisation.

The authorisation procedure for the approval of GM products in the EU is intended to ensure that the safety of these products is scientifically established before they are allowed on the market.

Directive 2001/18/EC on the Deliberate Release of GMOs into the Environment[2] came into force in October 2002 and determined the authorisation procedure for the approval of products consisting of or containing GMOs, before they may be placed on the market or released into the environment. According to these rules, a firm (whether local or foreign) had to submit an application for the approval of the relevant GMO or GM product to the Member State where it intended to first place the product on the market, together with an environmental risk assessment. If the Member State gave a favourable opinion, it informed other Member States of this. If there were no objections, the Member State could authorise the marketing of the product, in which case it could be marketed *throughout* the EU, provided it met all conditions attached to the approval. If other Member States did raise objections, a decision should have been taken on the issue at the EU level, on the basis of advice from the EU Scientific Committees composed of independent experts.

Products *produced from* GMOs, in other words products that are derived from GMOs but which no longer contain GM material (such as GM protein or DNA), are not covered by this Directive. For example, the process to make tomato sauce from GM tomatoes may eliminate all traces of GM protein or DNA from the final product. Such tomato sauce is not subject to the rules in this Directive.

Under Directive 90/220/EEC, which preceded Directive 2001/18/EC[3] and contained a similar authorisation procedure, 18 GMOs were authorised in the EU. However, no new authorisations have been granted since October 1998. Since the coming into force of Directive 2000/18/EC, 20 applications for authorisation have been received, including 7 which were pending under the previous Directive.

The *Regulation on Novel Foods and Novel Food Ingredients*[4] set out the authorisation procedure for novel foods, including GM food or food ingredients. Unlike Directive 2000/18/EC, this Regulation covered not only food and food ingredients consisting of or containing GMOs but also those *produced from* GMOs (i.e. where the GM material is no longer identifiable in the final product). The authorisation procedure under this Regulation was similar to that under Directive 2001/18/EC, except in the case of foods *produced from* GMOs but no longer containing GM material (such as highly-refined maize oil from GM maize). In the case of foods produced from GMOs, if such foods were substantially equivalent to their conventional counterparts (in terms of nutritional value, metabolism, composition, intended use and level of undesirable substances), there was no need to go through the authorisation procedure before placing those GM foods or food ingredients on the market. A notification to the Commission was sufficient, together with proof of substantial equivalence or an opinion of the authorities of a Member States to that effect. Two varieties of GM cottonseed oil, seven products produced from GM oilseed rape and four from GM maize have been placed on the EU market under this procedure for substantially equivalent products.

---

[2] *Directive of the European Parliament and the Council of the European Union on the Deliberate Release into the Environment of Genetically Modified Organisms and Repealing Council Directive 90/220/EEC,* (EC) No. 18/2001 OJ L 106/1 (Brussels: European Community, dated 12 March 2001).

[3] *Council Directive on the Deliberate Release into the Environment of Genetically Modified Organisms,* (EC) No. 90/220/EEC, OJ L 117/15 (Brussels: European Community, dated 23 April 1990).

[4] *Regulation of the European Parliament and of the Council concerning Novel Foods and Novel Food Ingredients* (EC) No. 258/97. OJ L 43/1 (Brussels: European Community, dated 27 January 1997).

Labelling.
Labelling rules were in place for all GM products that have been authorised for marketing in the EU. It is important to note that the purpose of labelling is not to indicate that the products are unsafe, but only to provide EU consumers with the information necessary to exercise their right to choose between GM and non-GM products. Unsafe GM products would not be authorised for marketing in the EU in the first place.

Directive 2001/18/EC requires that products containing or consisting of GMOs be labelled as such at all stages of placing on the market. The *Regulation on Novel Foods and Food Ingredients* required that GM food or food ingredients be labelled as such. If food was produced from GMOs but no longer contained GM material and was substantially equivalent to the conventional product, it need not be labelled. GM additives and flavourings are required to be labelled if GMO material is present in the final product, in terms of *Regulation 2000/50/EC* and GM seeds must be labelled as such under *Directive 98/95/EEC*. Until recently (see further at section 4) there was no EU legislation dealing specifically with labelling of GMO feed.

Small traces of GMOs may be present in non-GM food or feed due to accidental mixing during cultivation, harvesting, transportation or processing. This is known as "adventitious presence" of GM material. *Regulation 2000/49/EC*[5] provided that if the adventitious presence of GM material in non-GM food is below the threshold of 1%, labelling was not required, provided that the operators could prove that they had taken appropriate steps to avoid the presence of GM material as much as possible.

Traceability.
Directive 2001/18/EC contains a general obligation on Member States to ensure traceability at all stages of placing on the market of authorised products containing or consisting of GMOs. This requirement applied only to the notifier applying for authorisation for the GM product, not to later business operators on the marketing or distribution chain. This Directive did not define traceability or contain a coherent approach for its implementation.

3. The "De Facto Moratorium"

As seen from the discussion above the EU legislation on GMOs has never provided for a moratorium or ban on the authorisation of GMOs. Nevertheless, a "de facto" moratorium exists, as there have been no new authorisations of GMOs since October 1998. Certain Member States decided to block the authorisation process, due to concerns that the existing legislation is not sufficiently rigorous. These Member States insisted that additional rules on labelling and traceability were necessary to respond to consumer concerns.

In addition, certain Member States[6] invoked a "safeguard" clause in the Novel Food and Feed Directive and in environmental legislation. This clause allows them to temporarily ban already *authorised* GMOs in their territories if they have new information providing grounds for considering that the GMO in question presents a risk to human health or the environment. These decisions to temporarily ban authorised GMOs must subsequently be reviewed by the European Commission within 60 days, to

---

[5] *Commission Regulation Amending Council Regulation (EC) No 1139/98 Concerning the Compulsory Indication on the Labelling of Certain Foodstuffs Produced from Genetically Modified Organisms of Particulars Other Than Those Provided for in Directive 79/112/EEC,* (EC) No. 49/2000, OJ L 6/13 (Brussels: European Community, dated 10 January 2000).
[6] Italy, Austria, France, Germany, Luxembourg, Greece and the UK.

determine if they are well founded. Until recently, the Commission took no action in this regard, allowing these bans to remain unchecked. Recently, however, the Commission has officially requested seven Member States to repeal their safeguard measures, after the EU Scientific Committee(s) found that the bans were not justified on the basis of new information on the risks from the relevant GMOs. Decisions on the repeal of these measures are still pending.

The end of the *de facto* moratorium is, however, now in sight. On 28 January 2004, the EC Commission approved a proposal to authorise Syngenta's GM corn (Bt-11) for use in food, under the *Regulation on Novel Food and Feed*. If the Member States in the Council do not reach agreement regarding the authorisation of this product within 90 days, the Commission can grant the authorisation on its own initiative. This would be the first approval of a GM product in the EC since 1998.

4. New EU Legislation

In order to address the concerns of Member States and restore consumer confidence in the regulatory regime for GMOs sufficiently to lift the "de facto moratorium" and proceed with the approval of new varieties of GMOs, the EU adopted two new regulations, on 22 September 2003. These are *Regulation No 1830/2003 on Traceability and Labelling of Genetically Modified Organisms*[7] and *Regulation No 1829/2003 on Genetically Modified Food and Feed*.[8] Together with the now-amended Directive 2001/18/EC, these Regulations now form the regulatory framework for GMOs in the EU. This legislation must be complied with not only by business operators within the EU, but also by foreign traders wishing to export GM products into the EU. The relevant rules are discussed further below.

Traceability and Labelling
The new *Regulation on Traceability and Labelling of GMOs* establishes an EU system to ensure that all GM products (including food and feed) that have received EU authorisation to be placed on the market are traceable and labelled. This regulation applies not only to products consisting of or containing GMOs, but also to food and feed *produced from* GMOs (i.e. food/feed derived from GMOs but where DNA or protein of GM origin cannot be found in the final product). For example, highly refined soya oil produced from GM soya must meet the requirements of this regulation, even if the final product bears no trace of the GM material. However, the Regulation does not apply to products *produced with* GMOs. In other words, where GMOs were used as processing aids but did not form part of the product itself, the rules on traceability and labelling do not apply. For example, beef or milk from cows fed with GM feed or treated with GM veterinary drugs is not subject to the rules on traceability and labelling of this Regulation since the beef or milk itself is not genetically modified, although GMOs were used in the production process. Similarly, bread or cheese produced with the help of a GM enzyme does not need to be labelled.

The labelling rules in this Regulation require that pre-packaged products, intended for sale to the ultimate consumer, consisting of, containing GMOs must be labelled as follows: *"This product contains genetically modified organisms"*. In the case of bulk products consisting of or containing GMOs, the business operator must transmit information regarding the GMO content of the product to the operator receiving the product. These rules are intended to ensure that consumers' freedom of choice is protected.

---

[7] *Regulation of the European Parliament and of the Council Concerning the Traceability and Labelling of Genetically Modified Organisms and the Traceability of Food and Feed Products Produced from Genetically Modified Organisms and Amending Directive 2001/18/EC,* (EC) No. 1830/2003, OJ L 268/24 (Brussels: European Community, dated 22 September 2003).

[8] *Regulation of the European Parliament and of the Council on Genetically Modified Food and Feed,* (EC) No. 1829/2003, OJ L 268/1 (Brussels: European Community, dated 22 September 2003).

The rules on traceability are there to ensure that a product can be tracked throughout the production and distribution chains. Traceability ensures that it is possible to verify and control labelling, facilitates monitoring the potential effects of GMOs on the environment and enables withdrawal of a GM product if an unforeseen risk to health or the environment materialises. According to this Regulation, at each stage of the placing on the market of GMOs, business operators that use or handle these GMOs are obliged to retain specified information regarding the identity of the GMO that the product contains and the identity of the operator from whom and to whom the GM product has been made available, for five years after the transaction. This is known as the "one step forward, one step back" system. Thus, for example, a seed company that develops a GMO would have to provide the purchaser of the seed with specified information allowing the GMO to be precisely identified and would have to retain information on the identity of all the purchasers of this seed for 5 years. The farmer that buys the seed would have to retain the information on the identity of the GMO and the seed producer for 5 years and pass the information regarding the GMO on to the food companies that buy his harvest. The farmer would also have to retain information on the buyers of the harvest. The food company that buys and processes the harvest would have to retain information on the identity of the GMO and the farmer from whom the harvest was bought, for 5 years and transmit the information regarding the GMO to the supermarket that sells the processed food. The supermarket would have to retain this information for 5 years, but does not have to retain information on the identity of the final consumers that purchase the processed food product.

The traceability requirements differ with respect to products consisting of or containing GMOs on the one hand and products produced from GMOs on the other hand. Where products consist of or contain GMOs, the individual GMOs in the product must be traceable on the basis of authorised transformation events (in other words, the event by which a conventional organism is transformed into a GMO by the introduction of modified DNA sequences). Where products are produced from GMOs, there is no requirement of the identification of the GMOs from which they are produced.

It is envisaged that the European Commission will establish a system of assigning unique codes to GMOs, on the basis of the transformation event notified in the application for authorisation. It is the transformation event that ultimately determines the modified characteristics of the GMO. These codes will therefore assist in operationalising the traceability requirements in the Regulation.

GM Food and Feed
The *Regulation on GM Food and Feed* strengthens the rules for the authorisation, marketing and labelling of GM food and extends the rules to GM feed. It replaces the *Regulation on Novel Foods and Food Ingredients* with regard to GM food. It sets out a uniform authorisation procedure that applies to all food and feed consisting of, containing or produced from GMOs. It is important to note that no exception is made for "substantially equivalent" products. Just as is the case with the new *Regulation on Traceability and Labelling of GMOs*, the *Regulation on GM Food and Feed* does *not* apply to products *produced with* GMOs (i.e. where GMOs were used as processing aids but did not form part of the product itself), for example eggs from chickens fed with GM corn.

The authorisation procedure simplifies the procedure under Directive 2001/18/EC discussed above. Now the decisions on applications for authorisation are all taken at the EU level, rather than divided between the Member States and the EU. In addition, a single authorisation can be requested for both the release of a GMO into the environment and its use in food and/or feed. This is known as the "one door, one key" procedure. The scientific assessment of risks to health or the environment will be carried out by the European Food Safety Authority (EFSA), which includes Scientific Panels.[9] On this basis the European

---

[9]The Commission's scientific committees were transferred to EFSA in May 2003. Every reference in EC legislation to the scientific committees has been replaced by a reference to EFSA by Regulation 178/2000/EC. This regulation

Commission will propose the granting or refusal of authorisation. The Member States will decide on this proposal by a qualified majority vote. Products authorised under this regulation will be entered into a register of GM food and feed which will contain information on studies demonstrating the safety of the product and detection methods, which will have to be provided by the business operator applying for authorisation. Authorisation will be granted for a 10-year period, which is renewable with a new application.

With regard to labelling, this regulation requires that all food and feed consisting of, containing or produced from GMOs be labelled as such, regardless whether GM material can be identified in the final product. However, food or feed *produced with* GMO processing aids, such as eggs, milk or meat from animals fed with GM feed, need not be labelled.

Some GMOs have already been assessed as safe by the EFSA Scientific Panels but their final authorisation is still pending. In terms of the new Regulation, certain threshold tolerance levels are set for the accidental or adventitious presence of unauthorised GM material, *provided* that it has been assessed as risk free by the EFSA Scientific Panels. With regard to labelling, the adventitious presence of GM material in non-GM food or feed need not be labelled and traceable if it is below the threshold of 0.9 % and can be shown to be adventitious and technically unavoidable. With regard to authorisation, a tolerance threshold of 0.5 % has been established for the adventitious and technically unavoidable presence of unauthorised GM material in food or feed. Thus, no authorisation is needed if the GM material is below the 0.5 % threshold. This provision will expire after three years.

Co-existence

Co-existence refers to the ability of farmers to choose between growing conventional, organic and GM crops, in accordance with the relevant rules. The possibility of unintended admixture of crops could not only have great economic impact on producers but also affect consumer choice. In order for European consumers to have a real choice between organic, conventional, and GM products, it is necessary to ensure the ability of the agricultural industry to choose between the different forms of agriculture by preventing unintended admixture of crops. For this reason, the new Regulation allows EU Member States to take measures to ensure co-existence. In addition, on 23 July 2003, the Commission issued a recommendation with guidelines for the development of national strategies and best practices to ensure co-existence.[10] This recommendation is non-binding, and naturally applies only to the EU agricultural industry not to the agricultural industries of exporting countries.

5. Trade Implications

All businesses engaged in the production and trade of GM products for export into the EU will be affected by the new legislation. They will have to meet the new, often stricter, requirements contained therein.

Traders wanting to market products containing, consisting of or produced from GMOs in the EU, will have to apply for authorisation and ensure that their application is accompanied by all the required documents. Once authorisation has been granted, the business operator will have to ensure that the conditions attached to such authorisation are complied with. Once these conditions are met, the product

also created an EFSA Scientific Committee and eight Expert Panels. One of these is the Panel on GMOs. For more information on EFSA see www.efsa.eu.int.

[10] *Commission Recommendation on Guidelines for the Development of National Strategies and Best Practices to Ensure the Co-Existence of Genetically Modified Crops with Conventional and Organic Farming,* C(2003) (Brussels: European Community, dated 23 July 2003).

can be marketed in the EU for a 10-year period. Renewal of the authorisation can be applied for, for subsequent 10-year periods.

The new rules on labelling mean that products that are currently exempt from labelling requirements will now have to be labelled. Exporters will have to ensure that all products containing, consisting of or produced from GMOs are labelled as such, even if the presence of GM protein or DNA cannot be detected in the final product. However, an exception is made in the case of adventitious presence of GMOs in non-GM products below the 0.9 % threshold. Such products need not be labelled provided the relevant GMOs have been assessed as safe and their presence is technically unavoidable.

The traceability rules will mean that operators throughout the production and distribution chain will have to retain and transmit the required information regarding the genetic modification (transformation event). Operators that do not wish to label their products as genetically modified will have to establish segregation or identity preservation systems in order to supply non-GMO products. As mentioned above, provided the adventitious presence of GMOs in the non-GM products remain below the required threshold, no labelling will be required.

The new legislation also holds some benefits for exporters of GM products in to the EU. It increases legal certainty and transparency, so that they know in advance what will be required from them in order to market their products in the EU. It also builds confidence in EU consumers on the safety of GM products by responding to their concerns and providing an efficient system to ensure a high level of health and environmental protection from possible risks in GM products. This will contribute to overcoming the "de facto moratorium" on authorisations of GMOs so that it is once more possible to export GM products to the EU.

The new centralised authorisation procedure means that a single EU-wide authorisation request is sufficient. The procedure is also made more efficient and consistent, with specific deadlines to be respected by the Food Safety Authority and the Commission when issuing decisions.

The new traceability rules in place will ensure that the necessary information regarding genetic modification is available at all stages of the production and distribution chain, thereby facilitating accurate labelling without having to resort to costly sampling and testing procedures.

References (EC Legislation)

(1) Commission Recommendation on Guidelines for the Development of National Strategies and Best Practices to Ensure the Co-Existence of Genetically Modified Crops with Conventional and Organic Farming. C(2003) (Brussels: European Community, dated 23 July 2003).

(2) Commission Regulation Amending Council Regulation (EC) No 1139/98 Concerning the Compulsory Indication on the Labelling of Certain Foodstuffs Produced from Genetically Modified Organisms of Particulars Other Than Those Provided for in Directive 79/112/EEC. (EC) No. 49/2000, OJ L 6/13 (Brussels: European Community, dated 10 January 2000).

(3) Council Directive on the Deliberate Release into the Environment of Genetically Modified Organisms. (EC) No. 90/220/EEC, OJ L 117/15 (Brussels: European Community, dated 23 April 1990).

(4) Directive of the European Parliament and the Council of the European Union on the Deliberate Release into the Environment of Genetically Modified Organisms and Repealing Council Directive 90/220/EEC. (EC) No. 18/2001 OJ L 106/1 (Brussels: European Community, dated 12 March 2001).

(5) Regulation of the European Parliament and of the Council Concerning the Traceability and Labelling of Genetically Modified Organisms and the Traceability of Food and Feed Products Produced from Genetically Modified Organisms and Amending Directive 2001/18/EC. (EC) No. 1830/2003, OJ L 268/24 (Brussels: European Community, dated 22 September 2003).

(6) Regulation of the European Parliament and of the Council concerning Novel Foods and Novel Food Ingredients (EC) No. 258/97. OJ L 43/1 (Brussels: European Community, dated 27 January 1997).

(7) Regulation of the European Parliament and of the Council on Genetically Modified Food and Feed. (EC) No. 1829/2003, OJ L 268/1 (Brussels: European Community, dated 22 September 2003).

References

(1) Buonanno, Laurie, Sharon Zablotney and Richard Keefer, "Politics Versus Science in the Making of a New Regulatory Regime for Food in Europe." European Integration Online Papers 5, no. 12 (2001): 17.

(2) European Commission, European Legislative Framework for GMOs Is Now in Place. IP/03/1056 (Brussels: European Union, dated 22 July 2003), 4.

(3) Questions and Answers on the Regulation of GMOs in the EU. MEMO/02/160 - REV. (Brussels: European Community, dated 1 July 2003), 27.

(4) State of Play on GMO Authorisations under EU Law. MEMO/04/17 (Brussels: European Community, dated 28 January 2004)

(5) Mackenzie, Ruth and Silvia Francescon, "The Regulation of Genetically Modified Foods in the European Union: An Overview." New York University Environmental Law Journal 8 (2000): 530-555.

FOOD SAFETY: NOVEL FOOD, LABELLING AND MARKETING OF ALL GENETICALLY MODIFIED FEED AND FOOD

Kim-Helleberg Madsen
European Commission
Rue de la Loi
1000 Brussels, Belgium

This paper was presented orally, but was unfortunately not available at the time of publication.

RISK ANALYSIS FOR GMOs AND THE ROLE OF THE NEW EUROPEAN FOOD SAFETY AUTHORITY

Harry A. Kuiper
RIKILT- Institute of Food Safety.
Wageningen University and Research Center.
P.O. Box 230. NL-6700 AE Wageningen, the Netherlands

1.Introduction

Genetically modified (GM) crops were introduced commercially in 1996. Since that time the adoption of GM crops has continuously increased till a worldwide cultivated area of approximately 60 million hectares nowadays. Currently, cultivated GM crops carry traits that are predominantly of agronomic importance, e.g. herbicide and or insect resistance. Examples are soybeans, canola, cotton, maize, sugar beet and potatoes.

Several GM crops have been modified with traits that affect the functional and quality properties of the final product. Examples are long-ripening tomatoes with favorable post-harvest texture characteristics for processing into tomato paste, soybeans high in oleic acid and canola high in lauric acid. Moreover in the near future GM food crops will be developed with traits which may positively influence the nutritional and health status of consumers and animals. Well known examples are "Golden Rice" with pro-vitamin A introduced into kernels, and iron-fortified GM rice. The aim for these GM rice modifications is to alleviate vitamin A deficiency and/or anemia in developing countries where rice is a staple crop. These modifications have been accomplished through insertion of genes coding for entire non native metabolic pathways or through targeted alterations in existing ones. Other examples are high-expression levels of foreign or endogenous proteins with enhanced contents of essential amino acids (e.g high lysine corn), and GM plants designed as "protein factories" that serve as a matrix for functional or therapeutic proteins [Kok and Kuiper, 2003].

2. Safety Assessment Strategies for GM Crop Derived Foods

The safety assessment of GM foods is carried out through a comparison of the properties of the GM food with those of an existing food from which the GM food has been derived with a long history of safe use (Concept of "*Substantial Equivalence*"). The underlying assumption is that traditional crop plant varieties, although not elaborately tested in the laboratory prior to their marketing, can be considered safe and nutritious, notwithstanding the fact that many plant varieties may contain besides macro and micro nutrients, natural toxins. GM foods are thus evaluated for their safety in a *comparative* manner and no absolute safety can be warranted, given the inherent uncertainties associated with the assessment of traditionally produced foods.

Food safety evaluation issues of foods derived from GM crops comprise:
- Molecular characterisation of the introduced genetic fragment(s) as well as resulting new proteins or metabolites;
- A toxicological and nutritional assessment of newly introduced proteins or metabolites and of altered levels of endogenous compounds;

- Analysis of the composition of the relevant food plant parts with respect to key nutrients and anti-nutrients including natural toxins and potential allergens;
- Potential for gene transfer of specific genes from the GM food to micro-organisms in the human and animal gastro-intestinal tract;
- Potential allergenicity of the new gene products, or alteration of the intrinsic allergenicity of the modified food organism;
- Intake levels of the final product, including any altered constituent;
- Toxicity testing of the whole crop, or derived plant products, in cases where the composition of the whole crop has been changed significantly compared to the traditional counterpart or where there is a need to further investigate potential unintended side effects of the genetic modification.

Specific guidance on these issues has been provided by the Organization for Economic Co-operation and Development (OECD (1993), the Food and Agriculture Organization of the United Nations and the World Health Organization (FAO/WHO, 2000, 2001) and the Codex Alimentarius Commission (2003), and a detailed overview of safety assessment practises with GM food crops was published by Kuiper et al.(2001).

Application of the concept of Substantial Equivalence may result in identification of potential differences between the GM food and its counterpart, which are then further investigated with respect to their impact on human and animal health. Thus toxicological/nutritional testing is an essential part of the safety assessment paradigm for foods derived from GM crops. The concept of Substantial Equivalence is just a tool to identify potential differences, and part of a comprehensive comparative safety assessment approach. Recently it has been proposed to rephrase the concept of Substantial Equivalence into Comparative Safety Assessment Strategy (CSA) (Kok and Kuiper, 2003).

In order to standardise and harmonise the application of the concept of Substantial Equivalence, the OECD took up the challenge to formulate consensus documents on individual crop plants, including an overview of the key macro- and micro-nutrients as well as anti-nutritional factors, natural toxins, and their background values, when available, as reported in literature, for the different food crops (OECD 2003). Moreover a Crop Composition Database has been developed by ILSI USA which contains quality-controlled data and is a valuable supplement to the OECD consensus documents (ILSI 2003).

3. ENTRANSFOOD, The EU Funded Thematic Network on the Safety Assessment of Genetically Modified Food Crops

Despite intensive research efforts assessing the safety of GM crops, European consumers remain sceptical. Consumer and environmental organisations have voiced in particular concerns on the safety of these crops with respect to long-term effects on the environment and human health, and about consumers freedom of choice between GM containing and 'GM-free' foods. This debate demonstrates that a rigorous science-based risk assessment may not suffice to introduce a new food production technology into society, but that *societal* aspects should also be taken into account.

ENTRANSFOOD provided a platform for participants from a wide range of different perspectives and disciplines to interact and to explore the interdependence of scientific, regulatory and societal aspects of introducing GM food crops. Participants were recruited from academia, research centres, biotech and breeding companies, food industries, food retailers, regulatory agencies, and consumer groups across Europe. Forty-five Research Centres participated in 5 RTD projects, and 62 experts in 5 Working Groups (www.entransfood.com).

Main objectives of the Thematic Network were:

- to provide detailed guidance on how to perform the safety assessment of GM food crops;
- to develop strategies for the detection and assessment of unexpected effects possibly due to the genetic modification process;
- to assess the risks of transfer of recombinant DNA from GM crops to microbes or human cells;
- to examine the fate of GM raw materials and processed products throughout food production chains (traceability);
- to examine new strategies for the detection of GM raw materials, processed products and food ingredients;
- to examine societal aspects and consumers attitudes towards the introduction of foods derived from GM food crops;
- to establish a communication platform of producers of GM foods, scientists involved in food safety research and in societal aspects of GM food introduction, regulatory authorities, retailers and consumer groups.

3.1. Detailed Stepwise Procedure for the Safety Assessment of Foods derived from GM crops

The safety assessment procedure as designed by ENTRANSFOOD (Konig et al., 2004, in press) is conducted in four steps: (i) the description of the parent (recipient) crop; (ii) the description of the transformation process; (iii) the safety and allergenicity assessment of the gene products and metabolites; and (iv) the combined safety and nutritional assessment of the whole plant. The outcome of the safety evaluation is then combined with the intake assessment.

The parent crop should be characterised fully, together with close relatives, to understand any potential for toxicity, allergenicity, pharmacological, or anti-nutrient effects. The characterisation of the parent crop guides the choice of test parameters for the comparison of the GM crop to a close comparator, which is usually the non-modified parent crop.

The gene donor, transgenes and delivery process require a full description. This information, together with a detailed understanding of the function of any recombinant DNA sequences used in the transformation process, facilitates the hazard identification of the 'novel elements' that are transferred to the crop.

Newly expressed gene products (proteins and metabolites) should be thoroughly investigated using classical approaches for defined chemical substances. Knowledge of the amino acid sequence of the recombinant protein allows for screening of computer databases for any sequence similarities with known protein toxins and allergens. Repeated dose studies with recombinant proteins or derived substances are recommended to identify potential adverse long-term effects unless there is sufficient information to confirm the lack of toxicity or pharmacological activity of the recombinant proteins and metabolites.

Any significant differences in agronomic, physiological, and compositional characteristics between the GM crop and the conventional counter part are subject to further testing to assess potential health implications. The selection of the parameters for this comparison is guided by knowledge of the characteristics of the crop species that is transformed. Selected compositional parameters are representative of the main metabolic pathways in the plant and reflect potential consequences from the introduced trait; the assessment focuses on those that might affect human health, such as key nutrients, anti-nutrients, and allergens.

If the composition of a GM food crop is modified substantially or if there are any uncertainties on the equivalence of its composition to a traditional counterpart, the whole food derived from a GM crop should be tested, and dietary sub-chronic rat studies are recommended.

Current approaches for the assessment of the allergenic potential of novel gene products have been designed by FAO/WHO (2001) and adopted by the Codex Alimentarius Commission (2003). A weight of evidence approach is proposed that classifies proteins according to (i) the characteristics of the source of the recombinant protein, (ii) whether there is primary amino acid sequence similarity between the introduced protein and known protein allergens, (iii) whether specific binding to IgE antibodies occurs in *in-vitro* assays, and (iii) whether there is an indication for resistance to pepsin degradation. None of these single tests are conclusive and therefore all the available information must be taken into account.

3.2. Identification and Assessment of Unexpected Effects due to the Genetic Modification Process

The potential occurrence of unanticipated alterations in the composition of GM food crops as result of the genetic modification process is one of the key elements of the safety assessment procedure. Random insertion of genes into the genomic DNA of an host organism may besides intended effects, result in unexpected shifts in metabolic pathways leading to alterations in concentrations of nutrients and secondary metabolites or in theory even to the formation of new toxins. Unintended effects are known to occur in GM food crops, but are not unique for GM organisms, it happens frequently in conventional plant breeding via point mutations as well as chromosomal recombination mechanisms.

Possible alterations in the phenotype may be identified through a comparative analysis of growth performance, yield, disease resistance, composition etc. For spotting alterations in the composition of a GM organism compared to the parent, normally a *targeted* approach is used, i.e. measurements of *single known* compounds like macro and micro-nutrients, anti-nutrients and toxins. The targeted approach has its limitations with respect to a limited and 'biased' selection of compounds, while detection of unknown toxicants or anti-nutrients is not possible.

In order to increase the chances to detect unintended effects, profiling methods have been suggested as a tool for characterisation of changes in the composition of GM plants. This may be of particular relevance for GM food crops with improved nutritional or health protecting properties, obtained through the insertion of multiple genes. The *non-targeted* approach using DNA/RNA micro array technology, proteomics and hyphenated analytical techniques allows 'unbiased' profiling of possible changes in the physiology and metabolism of the modified host organism at different cellular integration levels. These techniques have recently been reviewed by Cellini et al.(2004, in press), Kuiper et al. (2003), and Kok and Kuiper (2003).

Detection of altered gene expression

Development of the microarray technology makes it possible to monitor the expression of thousands of different genes simultaneously and to link detected differences directly to the underlying gene(s). Research projects are ongoing to develop informative arrays for the tomato and potato as model systems for food plants (GMOCARE project, UK FSA project in progress). It is necessary to sample in a reproducible way comparable tissues of the plants that are in the same stage of development and have been grown under (near) identical environmental conditions. In order to evaluate the microarray fluorescent patterns adequately it will be necessary to gain sufficient insight into the *natural variation* in gene expression during different stages of development of the tissues of interest and under different environmental conditions.

Proteomics

The potential of proteomics to perform *comparative* analyses of protein patterns may be of importance for food safety assessment. The main approach currently applied involves two-dimensional gel electrophoresis (2-DE) followed by excision of protein spots from the gel, digestion into fragments by specific proteases, analysis by mass spectrometry and subsequent computer-assisted identification using databases. One of the major challenges is the quantification of proteins. At present, applicability of proteomic techniques is being studied within European multidisciplinary projects for the food safety evaluation of GM crops (GMOCARE, British FSA project).

Detection of differences related to the genetic modification might turn out to be very difficult, taking into account the large number of proteins not connected to such changes and given the natural variations in protein patterns due to different environmental conditions. The limited knowledge of natural variability of protein patterns of plants, demands the development of validated databases and further validation of the methods.

Metabolite profiling

Main approaches used for metabolite profiling are based on gas chromatography (GC), high performance liquid chromatography (HPLC), mass spectrometry (MS), nuclear magnetic resonance (NMR), or fourier-transform (near) infrared spectroscopy (FTIR). These methods are capable of detecting, identifying, resolving and quantifying a wide range of compounds in a single sample. Application is totally unbiased or targeted to metabolites in key metabolic pathways.

Data analysis

Application of profiling techniques even to a limited number of samples results in huge amounts of data. A meaningful analysis of profiles from a GM food and its non-GM counterpart with respect to safety implications, should be based on the entirety of potential differences, and multivariate techniques, *e.g.* principal component analysis or hierarchical cluster analysis are frequently applied. Moreover the set up of linked databases containing gene-expression, protein and metabolite profiles, reflecting different developmental stages and environmental conditions is essential. Furthermore standardisation of sampling procedures, and inter-laboratory testing and validation of these methods is needed.

3.3. The Risk of Gene Transfer from GM Crops

ENTRANSFOOD has evaluated the risks of horizontal gene transfer of recombinant DNA in foods derived from GM crops to microbes or human cells, and the impact of such a transfer event (van den Eede et al. 2004, in press). Gene transfer amongst different organisms is common in nature and has been a driving force in evolution. The relative risk of gene transfer of recombinant DNA from GM crops to microbes or human cells upon ingestion and its potential impact largely depend on two factors: the function of the transferred DNA in the recipient cell; and whether the recipient cell may have acquired the same gene from a source other than the GM crop.

The risks from gene transfer of GM crops that are currently commercial are deemed negligible. Transfer to microbes by transformation is a possibility but only consequential if trait expressed and confers selective advantage. Uptake of GM crop-derived DNA, including the transgenes by human cells of the gut or the immune system can not be ruled out; it is however very unlikely that transgenic DNA is stably integrated in somatic cells, or taken up in germ line cells. Even if it should be taken up, that trait conferred by the gene may not be expressed in human cells.

The risk of use of antibiotic resistance markers for selection of transformed plant cells should for instance be judged on a case-by-case basis, considering their frequency of occurrence in bacterial populations and the extent of clinical use of the antibiotics to which resistance is conferred. Some antibiotic resistance markers such as the *nptII* gene and the hygromycin resistance gene can be used without the risk of compromising human or animal health.

3.4. Detection, Traceability and Labelling

ENTRANSFOOD has also examined the conditions for proper detection, traceability and labelling systems (Miraglia et al., 2004, in press). Crucial for the development of any GMO detection and/or identification method is the availability of sequence information with relation to the genetic modification as well as the relevant reference materials. This information is needed for both EU-approved as well as unapproved GM varieties, requiring global exchange of information of GM events on the market and in development. An important step within Europe in this respect has been the establishment of the ENGL (European Network of GMO Laboratories), but much work remains to be done in a global perspective.

Appropriate traceability and segregation systems may reduce the necessity for stringent sampling schemes. The possibility to detect deviations from the documentation in traceability systems may, besides the administrative burden of the documentation itself, require additional measures such as storage of backup samples and additional testing. An appropriate traceability strategy for all GMOs in the food supply chain for the purpose of post-marketing surveillance will require new labelling and information transfer systems. It will be necessary to not only have the information on the label that GMO-derived materials have been used for the production of the food entity, but also which GM events composed the individual ingredients and to what extent. Traceability represents a valuable tool to face problems related to the introduction of GMOs in food and feed chain and to gain the confidence of the consumer toward this novel food. It is clear, however, that the implementation of any suitable system implies a substantial increase in overall cost of food production that will have to be absorbed by both producers and consumers.

3.5. Societal Aspects of the Introduction of Genetically Modified Foods

Societal aspects of the introduction of GM foods have been analysed by Frewer et al. (2004, in press). Much of the controversy associated with the commercialisation of GM foods has been the result of regulatory bodies failing to take account of the actual concerns of the public, which has fuelled public distrust in the motives of regulators, science and industry. In the case of GM foods, communication efforts have focused on adverse health effects whereas public concern has been much broader, focusing on risk (and risk perception), benefit and need.

The supposition that risk perceptions may be offset by perceptions of benefit has lead many scientists and industrialists to assume that if only a particularly desirable benefit can be developed in the context of GM foods then public acceptance will result. Problematically, how the public defines risk and benefit, and how the experts define the same issues, may be very different. Differences in perceptions of risk and benefit associated with various hazards exist between different countries and cultures, between different individuals within countries and within different individuals at different times and within different contexts.

If public confidence in science and technology is to be regained, it is important to explicitly incorporate public concerns into the risk analysis process, through developing new and influential methods of stakeholder involvement (including consumers) and consultation. Once public concerns are understood, they can be more effectively introduced into risk assessment and risk management practices.

4. Regulation of Food Safety in the European Union and the Role of the European Food Safety Authority

Recent food scares in Europe like the Bovine Spongiform Encephalopathy (BSE) and the dioxin crises have led to a dramatic decrease in trust of the general public in the safety of foods produced in Europe, and in the quality of existing food safety assessment and management procedures. Moreover concerns have been expressed by environmental and consumer groups with respect to the potential risks of the release of genetically modified organisms in the environment and their use as human food or animal feed. The European Commission has recognised these problems and addressed the issue of restoring the confidence of the consumers in the European food supply in the White Paper on Food Safety (EU 2000l). Consequently the Council of Agriculture Ministers adopted in January 2002 a Regulation setting up the European Food Safety Authority (EFSA) and a new framework for an EU Food Law (Regulation EC No 178/2002). The new legislation describes general objectives and principles of safety assurance of the food supply, including the precautionary principle, traceability, liability and protection of consumers interests.

4.1. Tasks and Structure of EFSA

The EFSA is the Authority that provides scientific advice and scientific and technical support for the Community's legislation and policies in the field of food and feed safety, and communicates on risks associated with the production of food and feed. The science of risk assessment of food and feed will be further developed at the highest standards and greater transparency of the risk analysis process will be pursued. EFSA is not involved and responsible for risk management of food safety related issues. EFSA consists of a Management Board, an Advisory Forum, has an Executive Director and staff and Scientific Panels and a Scientific Committee. The Management Board has 14 members appointed on the basis of their individual expertise and competence, and 1 member from the Commission (DG SANCO); 4 members have a background in organisations representing consumers or other interests in the food chain. The Advisory Forum consists of 15 representatives of the Member States from national food safety authorities and is chaired by the Executive Director; it advises the Executive Director and advises on scientific matters, priorities and the work programme and ensures close collaboration between national food safety authorities and the EFSA.

The EFSA Panels are focussed on risk assessment issues related to:
* Contaminants in the food chain
* Food additives, flavourings, processing aids and materials in contact with food
* Dietetic products, nutrition and allergies
* Biological hazards
* Additives and products or substances used in animal feed
* Genetically modified organisms
* Animal health and welfare

• Plant health, plant protection products and their residues

The Scientific Committee has a co-ordinating and pro-active role in further development and improvement of risk assessment models and strategies for food and feed related issues and in designing strategies to identify emerging risks. Tasks of EFSA are based on legal requirements or may come from the European Commission, the European Parliament, Member States or on its own initiative. EFSA's role in risk communication is to support the Commission with independent advice over food safety 'scares' and emerging risks, and a close co-ordination with Member States on public announcements on key EU-wide issues is foreseen in order to achieve a consistent and culturally sensitive output.

4.2. Regulation of Genetically Modified Organisms

The Directive 2001/18/EC (2001/18/EC) regulates the deliberate release into the environment of GMOs. It repeals the former Directive 90/220/EEC. The revised Directive strengthens the existing rules of the risk assessment and the decision-making process on the release of GMOs into the environment. It includes procedures and mandatory requirements for risk assessment. Public consultation is mandatory as well as the requirement for public registration of releases, and general rules on mandatory labelling and traceability at all stages of the placing on the market are given. Authorisations will be granted for a period of 10 years, subject to a post-market monitoring plan. Moreover the European Commission is required to report on the socio-economic advantages and disadvantages of GMOs which have been authorised for placing on the market.

The EU Regulation on Novel Foods and Novel Food Ingredients (EC/258/97) has until recently also governed the admission of GM foods on the market destined for human consumption. A new Regulation on genetically modified food and feed went into force September 2003 (EC 1829/2003). The new framework for risk assessment as laid down in the EU Food Law, is now applicable which implies that GM foods or feed should only be authorised for placing on the Community market after a scientific evaluation under the responsibility of the EFSA. Authorisation should not be granted for a single use, when a product is likely to be used both for food and feed purposes. Food and feed additives and flavourings, containing, consisting of or produced from GMOs fall also under the scope of the new Regulation. Processing aids derived from GM sources, only used during food or feed production, do not fall under the scope of the new Regulation, neither products obtained from animals fed with GM feed or treated with GM medicinal products.

A new Regulation has come into force in September 2003 which regulates the traceability and labelling of GMOs and the traceability of food and feed products derived from GMOs [EC 1830/2003]. All foods derived from GM crops, irrespective of whether DNA or protein of GM origin is detectable in the final product must be labelled. Traceability of GMOs should facilitate both the withdrawal of products where unforeseen adverse effects on human or animal health or the environment occur, and the targeting of monitoring to examine potential effects on the environment. Traceability should also facilitate the implementation of risk management measures in accordance with the precautionary principle. Information whether products contain or consist of GMOs and the unique codes for those GMOs at each stage of their placing on the market, provide the basis for appropriate traceability and labelling for GMOs. With respect to traceability, at the first and all subsequent stages of placing on the market of a product consisting of or containing GMOs, including bulk quantities, operators must ensure that information on whether the product contains or consists of GMOs, and has a unique identifier is transmitted to the receiver. For products consisting of or containing GMOs operators must ensure that pre-packaged and non- pre-packaged products are properly labelled.

Certain traces of GMOs may be adventitious or technically unavoidable, and need therefore no labelling and traceability requirements. For authorised GMOs a threshold of 0.9% is established. When the combined level of the adventitious presence of GMO traces is higher than the threshold, such presence should be indicated. For GM crops not approved in the EU a zero tolerance remains, unless they have already received a positive advice for marketing. In that case a threshold level of 0.5% will apply.

5. General Conclusions

1. Food safety assessment strategies for GM foods, using the Comparative Safety Assessment approach (concept of Substantial Equivalence), provide a high level of safety assurance that is at least comparable to that of conventionally produced foods. The strategy demands a thorough scrutiny of alterations in the GM food and their impact on human and animal health and the environment, compared to its traditional counterpart. Further harmonisation of the application of the comparative safety assessment approach is desirable.

2. ENTRANSFOOD has developed a detailed and step-wise procedure on how to carry out the safety assessment of foods derived from GM crops. This strategy is also applicable to the new generation of GM food crops with improved nutritional characteristics.

3. ENTRANSFOOD has developed and evaluated strategies to detect so-called unexpected effects in GM food crops. Existing *targeted* single compound analysis is considered sufficient to identify such effects, but *profiling* techniques, measuring many compounds simultaneously, may compliment this approach, once these methods are validated and databases have been established.

4. The risk of gene transfer from foods derived from GM crops that are currently commercial available has been evaluated by ENTRANSFOOD and is deemed negligible. The use of marker genes expressing resistance to antibiotics in use for medicinal or veterinary purposes has been evaluated. If the antibiotic is widely used or a tool of last resort, such genes should be avoided. Certain marker genes like *nptII* and the hygromycin resistance gene can be used without the risk of compromising human or animal health.

5. ENTRANSFOOD has examined criteria and requirements for the development of GMO detection methods. The availability of sequence information on the genetic modification and of relevant reference materials is crucial for the development of such methods. This requires joint efforts for global exchange of information on GM events on the market and under development. Appropriate traceability and segregation systems may reduce the necessity for stringent sampling schemes, but additional measures should be taken before administrative systems can effectively be applied.

6. Public concerns should be explicitly addressed and incorporated into the risk analysis process. Further research is needed on how to formalise public engagement and consultation into new working procedures and its impact on regulatory procedures and institutions involved.

7. The regulatory framework for GMO release in the environment and placing on the market within the European Union has thoroughly been renewed and is now in place. It includes specific regulations for environmental and food/feed risk assessment and requirements for post-market monitoring, as well as specific rules for traceability and labelling. This marks the beginning of a new era for the introduction of GM crops and derived foods and feed.

8. The establishment of the European Food Safety Authority (EFSA) is an important step forwards for improved, harmonised and transparent risk assessment and communication within the European Union, which may contribute to restore consumers confidence in the safety of the European food supply.

ACKNOWLEDGEMENTS

ENTRANSFOOD participants and Dr. Suzy Renckens (EFSA) are greatly acknowledged for their contributions.

References

(1) Cellini, F., Chesson, A., Colquhoun, I., Constable, A., Davies, H.V., Engel , K-H., Gatehouse , A.M.R., Kärenlampi , S., Kok, E.J., Leguay , J-J., Lehesranta , S., Noteborn , H.P.J.M., Pedersen, A.J., and Smith ,M. (2004) Unintended effects and their detection in genetically modified crops. In press.

(2) Codex Alimentarius Commission (2003), Codex Principles and Guidelines on Foods derived from Biotechnology, http://www.fao.org/english/newsroom/news/2003/20363-en.html

(3) Directive 2001/18/EC on the deliberate release into the environment of genetically modified organisms and repealing Council Directive 90/220/EEC. Off. J. Eur. Commun. L106/1-38

(4) Van den Eede, G., Aarts, H., Buhk, H.-J., Corthier, G., Flint, H.J., Hammes, W., Jacobsen, B., Midtvedt, T., van der Vossen, J., von Wright, A., Wackernagel, W., and Wilcks, A. (2004) The relevance of gene transfer to the safety of food and feed derived from GM-plants. In press.

(5) ENTRANSFOOD, European Network Safety Assessment of Genetically Modified Food Crops, www.ENTRANSFOOD.com

(6) EU (2000) White Paper on Food Safety, COM (1999) 719 final, 12 January 2000. European Commission, Brussels, Belgium. http://europa.eu.int/comm/dgs/health_consumer/library/pub/pub06_en.pdf

(7) FAO/WHO (2000): Safety aspects of genetically modified foods of plant origin. Report of a Joint FAO/WHO Expert Consultation on foods derived from biotechnology. World Health Organization, Geneva.

(8) FAO/WHO (2001) Allergenicity of genetically modified foods. Report of a Joint FAO/WHO Expert Consultation on Foods Derived from Biotechnology, Rome, 22-25 January 2001. Rome: Food and agriculture Organisation of the United Nations. http://www.fao.org/es/esn/gm/biotec-e.htm

(9) Frewer, L., Lassen, J. Kettlitz, B., Scholderer, J., Beekman, V., and Berdal, K.G. (2004), Societal aspects of genetically modified foods. In press.

(10) FSA project (Food Standards Agency) Transcriptome, proteome and metabolome analysis to detect unintended effects in genetically modified potato. http://www.foodstandards.gov.uk/science/research/NovelFoodsResearch/g02programme/g02project list/g02001/#re

(11) GMOCARE, European project New Methodologies for Assessing the Potential of Unintended Effects in Genetically Modified Food Crops, www.GMOCARE.com

(12) ILSI (2003), www.cropcomposition.org

(13) Kok, E.J. and Kuiper, H.A. (2003), Comparative safety assessment for biotech crops. Trends in Biotechnology 21, 10, 439- 444.

(14) König, A., Cockburn, A., Crevel, R.W.R., Debruyne, E., Grafstroem, R., Hammerling, U., Kimber, I., Knudsen, I., Kuiper, H.A., Peijnenburg, A.A.C.M., Penninks, A.H., Poulsen, M., Schauzu, M., and Wal, J.M. (2004). Assessment of the safety of foods derived from genetically modified (GM) crops. In press

(15) Kuiper, H.A., Kok, E.J. and Engel, K-H. (2003), Exploitation of molecular profiling techniques for GM food safety assessment. Current Opinion in Biotechnology 14, 238-243.

(16) Kuiper, H.A., Kleter, G.A., Noteborn, H.P.J.M., and Kok E.J. (2001), Assessment of the food safety issues related to genetically modified foods. The Plant Journal, 27(6), 503-528.

(17) Miraglia, M., Berdal, K.G., Brera, C., Corbisier, P., Holst-Jensen, A., Kok, E.J., Marvin, H.J.P., Schimmel, H., Rentsch, J., van Rie, J.P.P.F., and Zagon, J. (2004), Detection and traceability of genetically modified organisms in the food production chain. In press.

(18) OECD; Safety Evaluation of Foods Derived by Modern Biotechnology. Paris 1993: Organisation for Economic Co-operation and Development. http://www.oecd.org/dsti/sti/s_t/biotech/prod/modern.htm

(19) OECD (2003), Consensus Documents for the work on the safety of novel foods and feed. Paris, France: Task Force for the safety of novel foods and feed, Organization for Economic Co-operation and Development. www.oecd.org/oecd/pages/home/displaygeneral/0.3380.EN-document-530-nodirectorate-no-27-24778-32.00.html

(20) Regulation (EC) No 178/2002 of the European Parliament and of the Council of 28 January 2002 laying down the general principles and requirements of food law, establishing the European Food Safety Authority and laying down procedures in matters of food safety. Official Journal of the European Communities L31: 1-24. http://europa.eu.int/eur-lex/pri/en/oj/dat/2002/l_031/l_03120020201en00010024.pdf

(21) Regulation EC 1830/2003 concerning the traceability and labelling of genetically modified organisms and the traceability of food and feed products produced from genetically modified organisms and amending directive 2001/18/EC. Off. J. Eur. Commun. L 268/24-28.

(22) Regulation EC No 258/97 concerning novel foods and novel food ingredients. Off. J. Eur. Commun. L43/1-7.

(23) Regulation EC 1829/2003 on genetically modified food and feed. Off. J. Eur. Commun. L 268/1-23.

INTELLECTUAL PROPERTY RIGHTS REGIME IN THE EU: DIRECTIVE 98/44 ON THE LEGAL PROTECTION OF BIOTECHNOLOGICAL INVENTIONS

Jean-Luc GAL
European Commission.
Directorate General Internal Market. Industrial Property.
B - 1049 Brussels. Belgium

Abstract

This paper focus on the relevant EU provisions on patentability of plants and other related aspects such as questions raised by the Rio Convention on biological diversity. The aim of this paper is however not to deal with other controversial issues currently being discussed within the EU relating rather to the patentability of elements of the Human body (i.e. scope conferred to inventions related to human DNA sequences and patentability of Human Stem Cells and cell lines derived from them).

The introductory remarks will illustrate the long and thorough discussions which led to the adoption by the Council and the European Parliament of the directive 98/44[1] on the legal protection of biotechnological inventions. A brief overview will be given on the status of the implementation of the directive within the Member States.

The first part of the paper will deal with the relevant provisions of the directive relating to the protection of plants, animals and micro-organisms by patents. The directive lays down a very clear distinction between plant varieties which may be protected by a plant variety rights and plants which might be eligible for patent protection where the general requirements of patentability are fulfilled. It should be noted that this distinction is fully in line with article 27(3)b of the TRIPs Agreement. The European Court of justice gave some indications on this issue in its decision on the challenge by the Netherlands against the directive, which was rejected on October 2001. Moreover, a very important decision of the Enlarged Board of Appeal of the European Patent Organization (Novartis case) confirmed the interpretation given by the directive of this issue. Finally, the mechanism of cross licensing between plant variety rights and patents will be addressed.

The second part of this analysis will be devoted to the impact of international treaties on Industrial Property rights in the EC. First, it will be useful to assess the impact of the adoption of the FAO International Undertaking on Plant Genetic Resources on current patentable subject matter defined in the directive. Second, the main provisions of the directive which might have an impact on certain provisions of the Rio Convention such as the origin of the biological material used in a patented invention and the issue of Traditional Knowledge will be addressed. It should be borne in mind that the Communication of the EC and its Member States to the TRIPs Agreement has clarified the EC position on this issue. The latest developments at the international level will be presented: a very important proposal has been made by the Swiss delegation to the Patent Co-operation Agreement (PCT) reform working group. This proposal suggests in very practical terms that interested National

---

[1] Directive 98/44/EC of the European Parliament and of the Council of 6 July 1998 on the legal protection of biotechnological inventions, O J L 213 , 30/07/1998.

Patent Offices should be allowed to require certain information concerning the biological material contained in a patent application.

1. Introduction

Directive 98/44 on the legal protection of biotechnological inventions was finally adopted on 6 July 1998. An agreement was only reached after lengthy and thorough discussions – both within the Council and the European Parliament – lasting more than 10 years. It should be kept in mind that the first draft of the directive was rejected by the European Parliament on 1 March 1995. The rejection by the European Parliament was principally based on requirements of clarification between mere discoveries and inventions, in particular for the DNA sequences (deoxyribonucleic acid). The Commission took due account the concerns expressed by the European Parliament and the opinion of the Commission's Group of Advisers on 25 September 1996, which covered, in particular, ethical questions concerning genes.

The purpose of the directive was to help European biotechnological companies to become stronger *vis à vis* US and Japan competitors and to establish legal certainty in this area within the European Community. With regard to this last point, it was necessary to make clear that patents could be granted for inventions relating to biological material where criteria for patentability are met. This principle may apply to a large range of situations such as in particular plants, animals, micro-organisms and elements isolated from the human body.

According to Article 15(1) of the directive, deadline to transpose the directive expired on 30 July 2000. To date[2], only 14 Member States[3] have implemented Directive 98/44. A letter of formal notice was sent on 30 November 2000 to Member States who had not implemented the Directive on that date. On 19 December 2002, a reasoned opinion was addressed to Member States who had not implemented the Directive, which required a response by 19 February 2003. Preliminary comments were received from these Member States. On July 2003, an action against Member States having not transposed the directive was brought before the ECJ.

This paper will at first tackle the main provisions of the directive 98/44, in particular in connection with plants and animals. Indeed, due to the scope of the workshop to which this paper refers, the problems arising from the protection by patents of elements isolated from the human body will be excluded from further consideration. The second part of this paper will concentrate on the international developments which could have a significant impact on the application of the directive itself.

---

[2] The final version of this paper was prepared on the 7th June 2004
[3] Namely Denmark, Finland, United Kingdom, Ireland, Greece, Spain, Portugal, Sweden, Czech Republic, Slovakia, Poland, Hungary, Slovenia and Malta.

2. Main Provisions of the Directive

2.1 General Principle

The aim of the directive was to make clear that a patent could be obtained for an invention consisting of or containing biological material even if this biological material previously occurred in nature. This principle clearly states that inventions relating to biological material are not *per se* excluded from patentability. Patents can be granted for such invention where the patentability criteria (novelty, inventive step and industrial application) are fulfilled.

Before the adoption of the directive, the situation within the European Community was not crystal clear. Some countries and the European Patent Office granted patents for inventions relating to biological material while other countries were reluctant so to do. Now, one cannot argue that such an invention cannot be protected within the European Community.
This general principle can be applied to a large range of situations.

2.2 Protection by Patent of Inventions Relating to Plants, Animals and Micro-organisms

2.2.1 *Plants*
Article 4 of the directive lays down that plant varieties shall not be patentable. The reason for the exclusions of plant varieties from patentability can be explained by the existence of a specific protection for such varieties: the plant variety right which is provided for in Regulation 2100/94. This right complies with the *sui generis* protection set out in Article 27(3)b of the TRIPs Agreement[4].
In the light of this, Article 5(2) of Council Regulation (EC) 2100/94 of 27 July 1994 defines a plant variety as a plant grouping within a single botanical taxon of the lowest known rank[5].

The relevant provisions of the Directive which explain the differences between "plant variety" and "plant" are to be found in Article 4 and recitals 29 to 32.

On the basis of these recitals, it is clear that plant varieties are defined by their whole genome and are protected by plant variety rights. However, plant groupings of a higher taxonomic level than the variety, defined by a single gene and not by the whole genome, may be protected by a patent if the relevant invention incorporates only one gene and concerns a grouping wider than a single plant variety. A genetic modification of a specific plant variety is not patentable while modification of wider scope, concerning, for example, a species, may be protected by a patent[6].

The European Court of Justice in its decision dated on 7 October 2001 clearly reaffirmed these principles and dismissed the Netherlands' challenge against the directive[7].

---

[4] Agreement on Trade-Related Aspects of Intellectual Property Rights concluded in Marrakech in 1994.
[5] The taxon represents a group of organisms forming a clearly defined unit at each of the different hierarchical levels of classification.
[6] Points 44 and 45 of the judgment Case C-377/98 (see web page below).
[7] See:http://www.curia.eu.int/jurisp/cgi-bin/form.pl?lang=en&Submit=Submit&docrequire=alldocs&numaff=C-377%2F98&datefs=&datefe=&nomusuel=&domaine=&mots=&resmax=100

In the same vein, the enlarged Board of Appeal of the European Patent Organisation have considered a dispute involving this very problem[8]. Its decision is based *mutatis mutandis* on the same considerations as contained in Directive 98/44. First, the enlarged Board of Appeal emphasised that a claim wherein specific plant varieties are not individually claimed is not excluded from patentability under Article 53(b) EPC even though it may embrace plant varieties[9]. That Board added that if the subject of a patent relates, for example, to a process for obtaining plant variety, the rights conferred by that patent does not extend to the plant variety obtained directly by that process. Finally, the Board reaffirmed that the exception to patentability in Article 53(b), first half sentence, EPC applies to plant varieties irrespective of the way in which they were produced. Therefore, plant varieties containing genes introduced into an ancestral plant by a recombinant gene technology are excluded from patentability.

It should be noted that this distinction does not apply in the United States. The Supreme Court, in a Decision of 10 December 2001, judged that a patent could be granted for an invention relating to a plant variety if it met the conditions required (novelty, non-obvious subject matter, utility, sufficient description and deposit of biological material accessible to the public).[10] However, this kind of protection is in line with TRIPs Article 27(3)b, second sentence, which provides that the Members of this treaty shall provide for, the protection of the plant variety either by patents or by an effective *sui generis* protection or by any combination thereof.

2.2.2 Animals

The directive also excludes animal varieties from the patentability. However, there is no legal definition of an animal variety. This can be defined as a taxonomical grouping ranking next below a sub-species (where present) or species, whose members differ from others of the same species[11] or sub-species in minor but permanent or heritable characters[12].

The relevant provisions in the Directive are essentially Articles 4 and 6(2)(d). It should also be noted that there is no protection of animal varieties in Community law.

Under Article 4(1)(a), animal varieties are not patentable. However, inventions relating to animals are patentable if the technical feasibility of the invention is not confined to a particular plant or animal variety.[13] If an animal can be obtained only through genetic engineering, to the exclusion of any natural breeding, the invention relating to such an animal may be protected by patent.[14]

---

[8]   The Enlarged Board of Appeal of the European Patent Organisation, 20 December 1999, OJ EPO 3/2000, p.111.

[9]   Article 53(b) provides that European patents shall not be granted in respect of plants or animal varieties or essential biological processes for the production of plant or animals; this provision does not apply to microbiological processes or the products thereof.

[10] JEM AG Supply, Inc./Pioneer Hi-Bred International, Inc., 10 December 2001, BNA's 14-12-01 (Vol. 63, No 1552), p.144.

[11] A "species" is taken to mean a grouping of individuals with common morphological, anatomical, ecological, and ethnological, biochemical and physiological characteristics the individuals in which resemble each other more than they resemble other equivalent groupings. To belong the same species, the individuals must together have fertile common descendants in natural conditions.

[12] Definition taken from the Shorter Oxford English Dictionary, cf. the conclusions of Advocate-General Jacobs, point 131 (see above note 5 and6).

[13] It should be noted that, under Article 4(3) of the Directive, it is also possible to obtain a patent for technical processes which make it possible to obtain a new animal, or for the animal obtained by such processes.

[14] The reasoning put forward for plants applies *mutatis mutandis* to animals.

This question has been debated many times in the context of the patent for the Harvard 'oncomouse'. This patent relates to a mammal modified by genetic transfer. Thanks to this manipulation, the animal may, under certain conditions, develop tumours which can be used for cancer research.[15]

After more than 16 years of proceedings, the Opposition Division of the EPO responsible for this case decided on 7 November 2001 to limit this patent to transgenic rodents with the cancerous gene, and hence not to authorise its extension to all mammals with the introduced gene. In the United States, this patent was granted in its initial form, i.e. it covers any non-human transgenic mammal.[16]

In addition, Article 6(2)(d) lays down that processes for modifying the genetic identity of animals which are likely to cause them suffering, without any substantial medical benefit to man or animal, and also animals resulting from such processes, are excluded from patentability.
This exception is in line with the general exclusion concept laid down for invention whose commercial exploitation would be contrary to *ordre public* or morality.

Recital 45 states that the substantial medical benefit for man and animal referred to in Article 6(2)(d) must be present in the fields of research, prevention, diagnosis or therapy.

The European Group on Ethics in Science and New Technologies delivered an Opinion on 21 May 1996[17] on the ethical aspects of the genetic modification of animals. According to that Opinion, such modifications are admissible and can be patented, but in view of the consequences which the techniques used might have for human and animal health, as well as for the environment and society, extreme care is called for. This care must apply to both the obtaining of genetically modified animals and to their use and welfare.

To be exhaustive on this question, it should be pointed out that essentially biological processes for obtaining animals and plants are not patentable. *A contrario*, an essentially non-biological process will be patentable. It is for the courts to assess this difference.

2.2.3 Micro-organisms
The scope of the directive also covers microbiological processes. Moreover, again according to the TRIPs Agreement, micro-organisms must be patentable if the patentability conditions are met.
That is why Article 4(3) lays down that inventions which concern a microbiological or other technical process or a product obtained by means of such a process are not *per se* excluded from patentability.

2.3 Cross-licensing

The directive contains some provisions on cross-licensing between patents and plant variety rights. Article 12 provides that where a breeder or a patent holder cannot acquire or exploit a plant variety or a patent, they may apply for a compulsory licence for non-exclusive use of the plant variety right or the invention protected by the patent, subject to the payment of an appropriate royalty.

---

[15] For further information, consult the EPO website:
http://www.european-patent-office.org/news/pressrel/2001_11_05_e.htm
[16] US Patent 4.736.866.
[17] Opinion No 7 available at the following website: http://europa.eu.int/comm/european_group_ethics

Corresponding provisions have been inserted in the proposed regulation on the Community patent in order to deal with the dependency between national plant variety right and community plant variety rights on one side and national patents and the future community patent on the other. In addition, Article 29 of Regulation 2100/94 on Community plant variety right was amended by 29 April 2004[18] in order to put it in line with Article 11 of the directive. Indeed before the amendment of regulation 2100/94, this article only allowed a compulsory license to be obtained on a Community plant variety right for reasons of public interest.

To obtain such a license, the plaintiff has to demonstrate that he has applied unsuccessfully to the holder of the patent or of the plant variety right to obtain a contractual licence and that the inventions or the plant variety constitutes significant progress of considerable economic interest compared with the invention in the patent of the protected plant variety. It should be noted the conditions to applying for the grant of a license are identical *mutatis mutandis* to those existing in Article 31(l) of the TRIPs Agreement where patents are dependant each other.

2.4 Scope of Protection of Biotechnological Inventions

The directive does not intend to depart from the classical theory of patent law according to which a large scope of protection shall be granted. In this light, Article 8 (which covers also both process and product patent) of the directive states that any biological material derived from the initial biological material and which possesses the same specific characteristics as a result of the invention and obtained from through propagation or multiplication in an identical or divergent form is also covered by patent protection.

Article 9 adds that the protection conferred by a patent on a product containing or consisting of genetics information shall extend to all material in which the product is incorporated and in which the genetic information is contained and performs its function.

However, specific derogations are included in the directive. Thus, community exhaustion shall be applied. If a product relating to biotechnological inventions is placed on the market in the territory of one Member State by the holder of the patent or with his consent, he is not allowed to invoke the patent to oppose the free commercialisation of that particular product within the Internal Market. However, this exclusion does not extent to new biological material obtained from that non-protected material issued from other propagation or multiplication.

Article 11 which is inserted in the chapter dedicated to the scope of protection to be conferred to biotechnological inventions is an exception to the very broad scope of protection regarding such inventions. By derogation to article 8 and 9, the directive provides that the sale or other form of commercialisation of plant propagating material to a farmer by the holder of the patent or with his consent for agricultural use implies authorisation for the farmer to use the product of his harvest for propagation or multiplication by him on his own farm. This derogation called "farmer privilege" has its origin in the regulation on the UPOV Convention and at the European level in the Community plant variety right.

---

[18] Regulation (EC) No. 882/2004 of the European Parliament and of the Council of the 29 April 2004. OJ 30.4.2004, L162/38.

A same regime applies *mutatis mutandis* to animal variety, given that no community rights relating to animal variety is in force in the European Community.

3 International Dimension

The direction 98/44 may not be regarded as a piece of law independent from its international environment. Since the early 90's, international negotiators have been active in this area. Thus, Convention on Biological Diversity (hereinafter CBD), the TRIPs Agreement and the International treaty on Food and Plant Genetic Resources (hereinafter the International Treaty on PGRFA) were adopted respectively in 1992, 1994 and 2001.

Theses treaties have a clear influence on the interpretation of the provisions of the directive.

3.1 Relevant provisions of the directive

The directive has taken into account of the principles contained in the Convention on biodiversity. In this light, Article 1(2) clearly states that the directive shall be without prejudice to the obligations of the Member States pursuant to international agreements. A special emphasis is given on the necessity of compliance of the directive *vis a vis* the Convention on Biological Diversity. Thus recital 55 specifies that Member States must give particular weight to Article 3 (sovereignty on national resources), Article 8j (traditional knowledge), and Article 16 of the Convention (access to and transfer of technology) when bringing into force the laws implementing the directive.

Recital 27 of the directive appears to be of the utmost importance in so far as it set up an incentive aimed at encouraging the disclosure of the origin of the biological material used in a patent application, which is one of the cornerstones of the CBD. Although a recital, as such, does not empower any legal constraint, it remains a very good indication of the importance of the specific access and benefit sharing of the genetic resources in relation to biological inventions.

This recital was one of the more controversial issues discussed during the negotiation of the directive. Some countries strongly favoured a constraining character for this requirement. However, a majority of countries opposed a legal obligation while remaining aware of the importance of this provision.

Since the adoption of the directive in 1998, the adoption of the International Treaty on PGRFA and the ongoing discussion regarding the source of origin of biological material used in an invention have an impact on the way the directive should be interpreted.

3.2 The International Treaty on Plant Genetic Resources for Food and Agriculture

The International Treaty on PGRFA was adopted by the FAO (Food and Agriculture Organisation) conference on 3 November 2001. This treaty provides a legally binding global framework for the sustainable conservation of plant genetic resources for food and agriculture as well as a multilateral system[19] under which all parties to the Treaty not only have access to such resources but may also share the commercial and other benefits arising from their use.

The main relevant provisions of this treaty related to IP rights are Articles 12(3)(d) and 13(2)(d). Article 12(3)(d) provides that recipients (of the Multilateral System) shall not claim any intellectual property rights that limit the facilitated access to the plant genetic resources for food and agriculture, or their genetic part or components, in the form received from the Multilateral System. However, Article 3 paragraph 1 of Directive 98/44 clearly states that inventions which are new, which involve an inventive step and which are susceptible of industrial application shall be patentable even if they concern a product consisting or containing biological material or a process by means of which biological material is produced, processed or used. In addition, the second paragraph spells out that biological material which is isolated from its natural environment or produced by means of a technical process may be the subject of an invention even if it previously occurred in the nature.

Where an applicant has been able to isolate some biological material (for instance a DNA sequence) from its natural environment and develops an invention based on it, should that applicant be allowed to claim a patent for that invention without infringing article 12(3)(d) of the Treaty? It could be argued that the applicant, in many cases, has not obtained the biological material from the multilateral system and so may not be regarded as a recipient of such a system. However, it is not really obvious how such provisions can be interpreted.

For these reasons, the EC and its Member States made a declaration at the adoption of this treaty according to which the European Community interprets Article 12(3)(d) of the International Treaty on PGRFA as recognising that plant genetic resources for food and agriculture or their genetic parts or components which have undergone innovation may be the subject of intellectual property rights provided that the criteria to such rights are met.

An other article of this treaty is relevant for the IP rights: Article 13(2)(d) (ii). This article points out that a recipient who commercialises a product that is a plant genetic resource for food and agriculture and that incorporates material accessed from the Multilateral System shall pay to the system an equitable share of the benefits arising from the commercialisation of that product, except whenever such a product is available without restrictions to others for further research and breeding[20]. This article recognizes a difference of treatment between patent holder and breeder. Indeed, the exception according to which there is no obligation to pay an equitable share is clearly addressed to the holder of plant variety rights: according to the applicable legislation on Plant Variety Rights, any person who develops a new variety from an existing one may commercialise it without the consent of the holder of the first plant variety (except in the case of essential derived variety). In the context of the patent law, a research exemption is generally accepted for acts done privately and for non-commercial purposes or for acts done for experimental purposes relating to the subject matter of the patented invention.[21] However, any commercialisation of new plant variety comprising the patented invention requires the authorisation of the patent holder.

---

[19] All the resources contained in an annex attached to the Treaty are covered by the provisions of that Treaty.
[20] In which case the recipient who commercialises the genetic resources shall be encouraged to make such payment.
[21] See article 9 of the proposal.

3.3 The problem of the Source of Origin

This question, based on the substance of article 15 of the CBD, was first tackled by the Conference of the Parties (CoP) to the biodiversity Convention. The third CoP in Buenos Aires in November 1996 noted in Decision III/17 that 'further work is required to help develop a common appreciation of the relationship between intellectual property rights and the relevant provisions of the TRIPs Agreement and the Convention of the Biological Diversity, in particular on issues relating to technology transfer and conservation and sustainable use of biological diversity and the fair and equitable sharing of benefits arising out of the use of genetic resources, including the protection of knowledge, innovations and practices of indigenous and local communities embodying traditional lifestyles relevant for the conservation and sustainable use of biological diversity'.[22]

All the CoPs since the third one have tackled this problem. The last one in Kuala Lumpur, which was held in February 2004, reaffirmed its clear willingness to achieve significant progress on this issue.[23]

However, and in spite of the importance of decisions taken by the various CoPs, this paper will concentrate on the last developments of this issue within the context of the international Agencies relating to Intellectual Property.

3.3.1 The EC communication to the TRIPs Council

In September 2002, the European Community and its Member States tabled, in the context of the TRIPs Council, a proposal regarding the disclosure of the source of origin of the biological material used in a patent application.

The main characteristics of this proposal were as follows:

• The obligation should apply to all the WTO members; it should be kept in mind that some WTO Countries such as for instance USA, Japan and Canada are very reluctant *vis a vis* such mandatory system.
• All the patent applications based on or using biological material should contain such information.
• The obligation should not act as a formal or a substantial requirement. With regard to a formal requirement, an applicant is obliged to comply with all the different requirements of Offices for obtaining a patent. If the applicant does not provide one piece of information, he will be invited to provide the Office with this element under a specified time limit. If he complies with such obligation, the application will be prosecuted without any other additional proof. In the case where, he does not fulfil this requirement, his application is deemed to be withdrawn. In the context of a substantial requirement, the applicant has to justify that this specific requirement is fulfilled by providing the Office with reasonable proofs (it is the case for instance for the novelty, the inventive step, industrial application or the sufficient disclosure).

According to the EC proposal, in the case where the applicant does not mention the source of origin, the application would be nevertheless prosecuted.

The direct consequence of such proposal is the sanction for the non- or a false disclosure would be outside the ambit of the patent law. This proposal is clearly in line with recital 27 of Directive

---

[22] See the website of the CBD: http://www.biodiv.org/decisions/default.aspx?m=COP-03&id=7113&lg=0
[23] See decision VII/19: http://www.biodiv.org/decisions/default.aspx?m=COP-07&id=7756&lg=0

98/44 which clearly spells out that the non-disclosure could not lead to the revocation or the annulment of the patent application or of the patent itself.

Since this communication, no major progress has been achieved in the framework of the TRIPs Council. In addition, the failure of the last round of negotiation in Cancun in 2003 probably delayed the political impetus on this issue.

3.3.2 The Swiss proposal within the framework of WIPO

The Swiss proposal requires the applicant to indicate the origin of genetic resources which are directly incorporated in the claimed invention. In order to achieve this, the document proposed to amend rules 51.bis.1 and 4.17 Patent Co-operation Agreement (PCT).

It should be noted that this system was based on an "opting–out" concept which did not place any obligations on countries not wishing to impose such requirements. In addition, this system would apply to international applications in the national/regional phase or applications directly filed with the National Patent Offices.[24] The applicant would be obliged to disclose the origin of the source if known and, if it is not the case, (s)he would indicate that (s)he was not aware of it. The non-disclosure of the origin or incorrect disclosure could not lead to the non grant of the patent or the revocation or the annulment of the patent itself except where the source indicated or the non-disclosure by the applicant was considered fraudulent.

The advantage of the Swiss proposal lies in the fact that neither PCT nor PLT should be amended to allow such declaration. Indeed, only the rules of the PCT have to be changed. The major advantage of this situation is that no diplomatic conference need be convened for the introduction of this new system.
Finally, this proposal contains a clear incentive for disclosing the source of origin insofar as the applicant may do so at the international phase.

In addition, the Swiss proposal suggests that an Office receiving a patent application containing a declaration of the source of a genetic resource or knowledge should inform a government agency of the State declared as the source about the respective declaration. This agency could be the national focal point for access and benefit sharing. It should be noted that this proposal could lead to a modification of Article 30 of the PCT.

Practically speaking, if one representative State or Union of States such as the EC decided to introduce such a system, the source of origin of the biological material used in a patent application would be disclosed in more than 90% of all relevant applications![25]

---

[24] This last option is offered by PLT (Patent Law Treaty) which prescribed in Article 6§1: '(...) No Contracting Party shall require compliance with any requirements relating to the form or contents of an application different from or additional to:
(i) the requirements relating to form or contents which are provided for in respect of international applications under the Patent Cooperation Treaty;
(ii) the requirements relating to form or contents compliance with which, under the Patent Cooperation Treaty, may be required by the Office of, or acting for, any State party to that Treaty once the processing or examination of an international application, as referred to in Article 23 or 40 of the said Treaty, has started.'
[25] In practice, the EPO or/and the European Countries are generally designated in the PCT application. Moreover, since the first January 2004, all the Countries members of the PCT are deemed to be designated in the patent application.

Discussions regarding the Swiss proposal took place within WIPO at several meeting of the PCT reform Working Group (May 2003, November 2003 and May 2004).

Three main positions could be identified within the PCT reform WG:

- Countries which welcomed the Swiss proposal and wished to pursue discussions along the lines of this proposal (Switzerland of course, EC and its Member States, Norway, and probably China).
- Countries which are strongly opposed to the proposal (USA, Japan, Canada, Australia)
- Countries which thanked Switzerland for its proposal but considered that this proposal was too limited in its scope. (Brazil, India, Indonesia, African group)

The Working Group agreed to discuss the issue again at its next session on the basis of a revised proposal prepared by Switzerland. However, it appears likely that no significant success can be expected in the near future. Depending on the decisions of the next General Assembly of WIPO, which will take place in September 2004, the negotiation process could be boosted or delayed *sine die*.

4.   Conclusion

The merit of the directive is that it gives some guidance on the sensitive area of the biotechnological invention. It should be noted that it was not an easy task insofar as this field addresses ethical questions and complex processes that the European legislator is obliged to take into consideration. The directive as it stands proposes a balanced solution. The new international developments in the field of the genetic resources could have an impact on the directive even if the flexible provisions of that directive could embrace future steps in this area.

BIOTECH FOOD: IS IT SAFE ENOUGH OR IS SAFETY NOT ENOUGH? A NATIONAL CASE STUDY FROM THE NETHERLANDS

Jeanine A. G. van de Wiel
Safety Assessment of Novel Foods Unit
Health Council of the Netherlands
Parnassusplein 5. 2511 VX The Hague. The Netherlands

Abstract

With the application of biotechnology in agriculture - the green biotechnology - in Europe we find ourselves in a turbulent phase between the already more structured application of the red biotechnology (medicine) and the low profile application of the white biotechnology (industrial use of genetically modified micro-organisms). In economic terms this turbulent phase can be characterized as a million dollar/Euro World Trade Organisation WTO conflict. In terms of governance it can be characterized as controversies about facts and controversies about values. The application of this technology in agriculture has forced the European governments into a learning by doing policy cycle both at the national level and in the European Union but also in Codex Alimentarius and the United Nations Environment Programme.

In general the way out of policy controversies is creating a culture of consultation and dialogue about the topic. Has this been done with biotechnology in agriculture? At first it was not such a big issue. There was a general food, feed and environmental safety check by national implementation of a European Directive of 1990. Then food safety became an issue in Europe. In 1997 the European Regulation concerning novel foods and novel food ingredients entered into force. From then on manufacturers of novel foods were required to submit a complete safety dossier before they are allowed to launch these products onto the market. The pre-market safety assessment is putting into practice the precautionary principle, a major cornerstone of European food safety policy. All European countries got a Food Assessment body to do this task. In the Netherlands this is the Health Council Committee on the safety assessment of novel foods. The safety assessment involves the evaluation of data on chemical composition of the food, nutritional value, biomolecular characterization of the source crop and on toxicological and epidemiological studies. A food is regarded as safe as there is proven absence of all adverse characteristics that are detectable by means of current knowledge. All advisory reports to the Ministers of Health and Agriculture are made public on the website of the Health Council. If European member states do not agree the European Scientific Committee on Food re-evaluates the dossier, the first safety assessment and additional data from the company. These reports are also made public. Although at least three crops were judged safe at both the level of a national Food Assessment Body and the European Scientific Committee on Food there were no new GM crops allowed to the European market for four years. Partly this was because preconditions for risk management where not yet fulfilled (traceability, reference material, detection methods) and maybe because there was not yet enough trust in the expert judgement. But the largest hurdles were that, meanwhile, the decline of biodiversity became an issue and the freedom of food choice for the consumer and the requirement of coexistence with other types of agriculture and the fear of monopolization of biological resources.

This means that the single crop related evaluation of food safety, feed safety and environmental safety that was practised so far does not fully comply with European countries needs. It is a challenge to realize a broader agricultural biotechnology assessment. This could be done by adding an ethical/societal evaluation panel to the yet existing food and feed safety evaluation panel and environmental risk evaluation panel. Room for this was already created in existing food and environmental legislation, guidelines and protocols at the international, the European and the national level.

Purpose

Agricultural biotechnology raised a conflict within Europe and a trade war between Europe and the USA. In Europe the food safety case is nearly solved. By studying the ten years debate on GM food safety probably we can learn to facilitate and improve the discussion about other more complex issues in relation to agricultural biotechnology.

1. Introduction

With the application of biotechnology in agriculture - the green biotechnology - in Europe we find ourselves in a turbulent phase between the already more structured application of the red biotechnology (medicine) and the low profile application of the white biotechnology (industrial use of genetically modified micro-organisms). In economic terms this turbulent phase can be characterized as a million dollar/euro WTO conflict. In terms of governance it can be characterized as controversies about facts and controversies about values. The application of this technology in agriculture has forced the European governments into a learning by doing policy cycle both at the national level and in the European Union but also in Codex Alimentarius and the United Nations Environment Programme.

In general the way out of policy controversies on a strategic level is creating a culture of consultation and dialogue about the topic (GR96). Has this been done with biotechnology in agriculture? At first it was not an issue. There was a general food, feed and environmental safety check by national implementation of a European Directive of 1990. Then food safety became a big issue in Europe. One of the reasons was BSE. Beneath the surface there was the more general uncertainty about how to feed oneself in a technological culture (Kort01).

2. Safety and Precaution

In 1997 the European Regulation concerning novel foods and novel food ingredients entered into force (EC97).

Prior to 1997, foods were not subject to systematic government safety assessments before they entered the European market[1]. Since May 1997, however, food producers have been required to submit a complete safety dossier before the product in question can be introduced to the market. Once approval has been granted, the product can be placed on any market within the European Union. However, the product is not admitted until the assessing body is fully satisfied that it is safe. This provides more room for precaution than did the previous situation. Then, the government was only able to intervene if there was hard evidence of adverse health effects. The European White Paper on Food Safety, which was published in the year 2000, placed further emphasis on the principle of precaution (EC00). The pre-market safety assessment is one of the ways in which the principle of precaution has been made operational. The precautionary principle is a major corner stone of European food safety policy.

3. Transparency in Expert Consultation

In a request for advice to the President of the Health Council, the Netherlands Minister of Public Health, Welfare and Sports wrote that her policy was intended to achieve the greatest possible openness and transparency with regard to the assessment procedure for novel foods. In February 1999, the Health Council set up the Committee on Safety Assessment of Novel Foods (VNV Committee). The Committee

---

[1] For several years, however, the United Kingdom and the Netherlands each had appropriate national regulations

consists of nutritional scientists, toxicologists, epidemiologists, molecular biologists, clinicians and a statistician.

During the past five years, the Committee has gained experience in assessing the safety of a wide variety of novel foods. It aims on consistency and predictability in the requirements that are imposed on dossiers. In 2002, in a general framework report it is indicated, for a variety of product categories, the minimum study data that is required to perform a safety assessment. It is also shown how the Committee tackles such an assessment. The VNV Committee's recommendations can be directly accessed via the Health Council's website. The information available there shows which dossiers are currently being processed (www.healthcouncil.nl/vnv.php). Dossiers can be inspected on request, at the Health Council's library. The individual members of the Committee notify whenever they have an interest that might influence them in reaching a verdict on given dossiers. In such instances they take no further part in the safety evaluation.

4. Quality in Expert Consultation

The VNV Committee makes its assessment on the basis of the current scientific state of affairs. It is checked whether the data supplied by the manufacturer are complete and accurate and whether it concurs with the manufacturer's conclusions ('first assessment'). The Council draws up a report of its findings and submits this to the minister. The minister raises the Dutch conclusions on a food in consultations at European level, in the Standing Committee on the Food Chain and Animal Health. All European member states are invited to give a so-called second opinion about the dossier and about the first assessment before the Standing Committee passes judgement. If a dossier raises numerous questions then the European Commission requests advice from the Scientific Committee for Human Nutrition. If there is still no agreement then a decision is taken by the European Council of Ministers.

In the case of a product that is essentially equivalent to another product, a notification of market introduction is sufficient. This means that one of the Member States checks the notification dossier, if this is found positive the product's introduction to the market can go ahead. The comparative approach in general, and the associated concept of 'substantial equivalence' in particular, has given rise to a great deal of confusion. In the European Union, a food that one of the member states considers to be 'substantially equivalent', can immediately be placed on the market via a so-called notification (EC97). This happened, for example, in the case of highly refined products such as soya bean oil or rapeseed oil that have been derived from certain genetically modified crops. They contain no demonstrable amounts of protein and, in terms of their composition (as determined by chemical analysis) and nutritional value, they are identical to the oils produced by conventional crops. Accordingly, these can be placed on the market after the consent of one Member State.

The disadvantage of this approach, however, is that the other states are unable to inspect the dossier and so are unaware of the changes in the crop's molecular biology. Another unknown factor is the risk of unplanned adverse changes in composition that might slip through undetected because the chemical analysis procedures used had targeted other components. The Netherlands was quick to advocate the full assessment of all products that result from genetic modification, in the interests of openness and transparency. This is also to become current practice within Europe with the new regulation.
For those assessing the safety of novel foods, there is no protocol setting out exactly which studies must be conducted and how their results should be interpreted. It has been decided that a case by case, step by

step approach should be adopted[2]. In this connection, the safety assessment of novel foods gives rise to a number of scientific challenges. The assessment involves five types of study. These correspond to the disciplines of analytical chemistry, nutrition, molecular biology, toxicology and epidemiology. The safety dossier is structured in accordance with the relevant European Commission recommendations (EC97a).

The themes that are covered are based on reports from various bodies working in the area of novel foods, namely the WHO/FAO (WHO91. FAO96) and the OECD (OECD93, OECD96). The Health Council has also considered this topic in the past (GR92), as has the Food and Nutrition Council (VR93).

Since the publication of the EC recommendations, work has been done at international level (FAO01, SCF99, SSC99, OECD98, OECD00, WHO00) to make them more explicit and to adapt them to the scientific situation. Principles and guidelines have been drawn up in the Codex Alimentarius. These are valid throughout the world for the safety assessment of foods from genetically modified sources (CA03).

The safety dossier is submitted by the applicant. This is why the dossiers are subject to strict quality rules. Studies should have been conducted to generate data for safety assessments. Their design and implementation must comply with sound scientific principles and guidelines, and with Good Laboratory Practice (where applicable). Primary data have to be delivered to the assessing institute upon request. They must have been obtained by scientific methods and analysed using appropriate statistical techniques. The sensitivity of the analytical techniques used must be documented.

The results of the various types of studies are integrated into a well-considered final verdict. The conclusion that a product is safe is based on the proven absence of all adverse characteristics that are detectable by means of current scientific knowledge, and on the extent to which the product corresponds to well-known, 'traditional' foods which are seen as safe.

The comparison of a genetically modified crop with its conventional counterpart involves an investigation of the crop's composition. Within the OECD, there is a debate about which components of each agricultural crop should be examined. The VNV Committee places emphasis on the measurement of secondary plant metabolites as one of the tools to check if unintended effects were generated with the genetic modification, for instance disturbance of normal physiology. In 2002 the Committee (in collaboration with America's FDA and the RIKILT - Institute of Food Safety) has drawn up an OECD consensus document on maize (Env02). This document identifies the nutrients, anti nutrients, toxicant and secondary plant metabolites that are useful to measure in the course of a safety assessment of genetically modified maize. It provides details of the concentration ranges of components in conventional maize lines. Other countries drew up consensus documents about rapeseed, soya, potatoes, sugar beet and wheat. Future documents are planned for rice, sunflowers and hops. The use of analytical chemistry to measure various secondary plant substances can be seen as a forerunner of so-called profiling methods. The latter are techniques for elucidating a plant's metabolic pathways and for changing them by varying the conditions under which growth and development take place, or by the use of genetic modification.

Data on the composition of a genetically modified plant and that of the most relevant reference plant should be obtained from a series of different sites, all of which are representative of those areas in which the variety in question will be commercially cultivated. The deciding factor here must be the site of commercial cultivation. Given the huge range of possible variation in conditions (soil, type of weather),

---

[2] This approach is currently being applied in other areas, see 'Toxicology testing: a more efficient approach' Health Council 2001/24.

it is not possible to make an absolute statement regarding the requisite number of sites on the basis of theory alone. It is recommended that cultivation should be continued for several seasons at each site. At each location, sufficient sub-experiments should be included to allow a mean, a measure of dispersion and a range to be established at each site for each specific component in both the genetically modified plant and the control plant.

Applicants should interpret statistically significant differences between genetically modified and control plants in relation to the observed variation between sites, thereby enabling a comparison to be made with data from the literature.

Sufficient information is usually available concerning the intake and frequency of use of agricultural crops, although maize and soya fractions for example have an extremely wide and varied area of application. Induced changes in the present generation of genetically modified agricultural crops are usually of only agronomic importance. They are unlikely to bring about any changes in the current patterns of consumption. Intake data are important in determining the difference between the tested doses of the new protein and actual consumption. The manufacturer must make an estimate of the consumer's total exposure to the new protein. The estimated level of exposure is then compared against data from acute toxicity tests on the protein and on subchronic experimental animal research using the whole soya grain or cereal grain.

Molecular biology techniques can be used to identify the differences between the DNA of the transgenic plant and that of the conventional line. The data of fundamental importance here is the base sequence of the newly inserted DNA and that of the flanking DNA in the transgenic plant. Furthermore, the way in which the transformation was carried out is also of importance, as is the expression of the new genes and whether their integration into the genome can be anticipated to have effects on the plant's original genes. In addition, molecular biology techniques are used to demonstrate that the new DNA maintains a stable presence through succeeding generations.

On the basis of its experience in the assessment of genetically modified crops, the VNV Committee concludes that unplanned changes often occur in such plants. This can result, for example, from incomplete insertion of the new DNA in question or from the duplication of certain parts of it. Additional portions of the (plasmid) DNA that was amplified in bacteria are occasionally transferred. Alternatively, there may be extra rearrangements of the plant's DNA. This need not necessarily lead to a product that is less safe, however it does require that an accurate pre-market assessment is carried out so that a properly supported safety verdict can be issued. It must be kept in mind that the classical methods of plant improvement can also be associated with unplanned changes in the plant's DNA.

All possible translation products are compared to databases of allergens and toxins, and to databases containing information on all known proteins that are known to have a particular function.
In the genetically modified plant, the new protein that provides the novel and desirable property in question (such as herbicide tolerance or protection against damage by insects) is tested for acute toxicity in an experimental animal. Testing for allergenic characteristics is also carried out. An important aspect here is the degree to which the protein is broken down (digested) by the gastric juices. If such breakdown occurs rapidly, then there is only a small risk of allergenic activity. In addition, the composition of the protein is compared to that of known allergenic proteins. If the protein does not correspond to any known allergen and if it breaks down rapidly in the stomach, then it is considered to be safe.

Proposals have been put forward for a more extensive test protocol for allergenicity (FAO01). One problem with this approach is that the use of human serum for routine screening is not feasible in practice, another is that no sufficiently validated experimental animal model is yet available. The

Committee feels that the current approach to the pre-market assessment of the first generation of genetically modified crops is satisfactory (Tay01).

The Committee has introduced a new requirement since it became clear that the molecular biology data and the compositional analysis do not provide a complete picture of the genetically modified crop. This new requirement is for a subchronic study in rats, in accordance with the OECD 408 protocol. The crop plants that are most often used for genetic modifications are maize and soya. These crops are used in the standard diet given to experimental animals (the so-called 'lab chow') at concentrations of up to 30%. This means that there are historic controls for (conventional) maize and soya. When testing genetically modified crops, the conventional line and the modified line can be compared to the historical data.

To date, there have been no human studies of genetically modified foods in the pre-market stage of investigation. With regard to the genetically modified foods that have been submitted to date, the Committee considers that the entire package of studies (molecular biology, analytical chemistry, nutritional and toxicological) provides an adequate foundation on which to base a verdict concerning the product's safety for consumers.

5. Future Products and Future Procedures

To date, the genetically modified products examined by the VNV Committee have been almost entirely limited to those that are intended to provide improved cultivation. The Committee predicts that, in the near future, they will also receive requests for crops with combinations of characteristics, such as resistance to insect pests and herbicide tolerance. One example of a product in which changes have deliberately been made to the composition of the end-product is the soya bean with an elevated oleic acid content. Manufacturers can be expected to develop more products like this.

One example of a product that is not of direct importance to the Netherlands, but which is indicative of current developments, is genetically modified rice with an elevated carotene level. Some are convinced that this can solve a major health problem in a number of countries, but many questions still remain to be answered. For example, it is not known whether the long-term use of this product really will result in an improvement in people's vitamin-A status. Tackling one problem may well cause deficiencies to develop in another area. In view of the potential health gains, several countries have decided to grant approval to the rice and to adopt a 'wait and see' attitude with regard to the outcome.

Other future products will be derived from genetically modified animals.
Scientific developments in the field of genomics will have impact on the safety assessment procedure. Once techniques such as transcriptomics, proteomics and metabolomics have been further developed, they can be used to replace part of the molecular biological analyses, compositional analyses and experimental animal studies that currently make up part of the safety dossier. This will enable savings to be achieved in terms of cost, time and experimental animals. Another advantage is that multiple gene products can be analysed at one go, when comparing the modified plant to its conventional counterpart. Another beneficial development is that more fundamental information will become available about the genome of economically important crops. This in turn will make it possible to produce better predictions concerning possible effects associated with the insertion site of the new DNA. Producers may be able to use such knowledge to insert new DNA in a more controlled manner, or to select specific transformed organisms.

The Commission notes that the pre-market safety assessment for novel foods has been worked out in much greater detail than the monitoring of products that have already been placed on the market.

Commission Recommendation 97/618/EC makes several references to market monitoring as a valuable option in the safety assessment of novel foods. However, this recommendation contains no guidelines on how this approach should be implemented. Market monitoring can be used to analyse the consumption of foods in a real situation. Data on actual consumption is considered to be important if the pre-market study includes a restriction regarding the level of consumption or consumption by specific groups. In addition, market monitoring provides an opportunity to study the consequences of consumption in a large group of people that is representative of the entire population. This makes it possible to conduct research into more long-term use, uncommon effects and changes in patterns of consumption caused by novel foods.

While there is indeed a social demand for market monitoring (for example, in the case of foods that result from genetic modification) the Committee is unsure whether it is possible to do this within the framework of a sound scientific design. This means that the system should not generate too many false positive or false negative signals. This can be investigated by means of a large-scale, long-term study. The Committee proposes the use of a food monitoring system with four pivotal points. The first such point would be a government-supported complaints line for all consumer complaints associated with health and foods. This would enable any side effects produced by a product to be traced, provided that they arise soon after consumption and that they attract attention. One such example is food allergies. The second pivotal point in the food monitoring system is a continual monitoring of consumption data for foodstuffs, carried out jointly by government and industry. A precondition here is that this must facilitate a detailed breakdown of data on individual products into information on their ingredients (down to the molecular biology level, in the case of GMOs).

The third pivotal point is long-term epidemiological, prospective, cohort studies into the relationship between chronic diseases and diet. The fourth pivotal point is an active market monitoring programme, carried out by companies, for novel foods that contain bio-active ingredients. The aim here is to check the accuracy of the presumed (safe) intake by the target group.

European legislation in the area of novel foods is in a state of flux. A proposal for a separate regulation for genetically modified food and cattle feed is adopted recently. A proposal for more effective traceability and labelling (EC01, EC01a) is also accepted. The European environmental-approval procedure has already been tightened-up (EC01b).

The new European Regulations concerning genetically modified foods demand details of the ways in which these can be traced in the food chain. This will be used to check the labelling, thus safeguarding the consumer's freedom of choice. From the point of view of food safety, traceability is useful for market monitoring. In the present situation, whenever there is a food incident in which public health may be at issue, an investigation is carried out to identify those products that contain the ingredient in question. This process can be accelerated by having immediate access to a database containing a summary of all end-products, together with their respective genetically modified ingredients. If an up-to-date database is available, it is immediately clear whether, for example, a cluster of allergenic symptoms that is associated with certain foodstuffs also involves the consumption of new types of proteins by consumers.

As yet, however, there is no systematic summary of specific ingredients from genetically modified sources and their uses in consumer products. Should incidents occur, like that involving StarLink maize in the United States, it is not clear whether the industry will be able to make this information available quickly. Compliance with the European labelling and traceability proposal will mean that it must be possible to trace all genetically modified crops, from the field to the end product.

6. Beyond Food Safety

Although at least two crops were judged safe at both the level of a national Food Assessment Body and the European Scientific Committee on Food there were no new GM crops allowed to the European market for four years. Partly this was because preconditions for risk management where not yet fulfilled (traceability, reference material, detection methods) and maybe because certain advocacy groups had not yet enough trust in the expert judgment. However, the largest hurdles were that meanwhile the decline of biodiversity became an issue as well as the freedom of food choice for the consumer, the requirement of coexistence with other types of agriculture and the fear of monopolization of biological resources. These are very heterogeneous themes. The discussion is about facts but also largely about values. How can the agricultural biotechnology assessment be structured to include all these themes? Indeed in the field of medical technology assessment within the Health Council there is experience with evaluating safety, efficacy, societal and ethical implications. Evaluation of societal and ethical implications makes use of four basic ethical principles. These four principles are to do no harm, to do well, to realize a proportionate distribution of advantages and disadvantages of the application evaluated (be fair) and to respect the autonomy of the individual. A broad risk assessment of biotechnology applications in agriculture could be structured along these lines.

In the Netherlands, the Minister of Housing, Environment and Spatial Planning asked his advisory body in the field of gmo's, the Committee on Genetic Modification, (COGEM) to develop an integrated ethical testing system for genetically modified plants and animals. In a very recent report the COGEM proposes to introduce the fairness principle in the decision making. They propose to add a cost/benefit form to every gmo dossier to be filled in by the applicant. The purpose is to describe all envisioned harmful effects but also all provisioned benefits and interprete these for the concrete casus in its context (COGEM03) to reach a final decision on release.

In addition to a safety test, foodstuffs with specific bio-active components may also be subject to an efficacy test. The Minister of Health, Welfare and Sport has asked the Health Council to explore this option. Recently a set of scientific assessment criteria was drawn (GR03) and some thoughts were given to ethical dilemma's that accompany the introduction of foods with specific bio-active compounds.

7. Conclusions and recommendations

By analyzing the ten year debate about food safety it can be seen that in Europe it was difficult to get an operational consensus on the common value of food safety in relation to agricultural biotechnology.

Four discernible factors contributed to find a workable solution: make the expert consultation procedure(s) very transparent, improve the quality of the expert judgement and enlarge it to unintended effects. Further, not only create EU regulation on risk assessment but also structure and facilitate risk management to some level and use local (national) expertise to provide know-how and initiatives for handling options in delicate and controversial issues that accompany this new technology.

A novel recommendation upon which there is not much experience yet is to realize a broad agricultural biotechnology assessment. This can be done by adding an ethical/societal evaluation panel to the yet existing food and feed safety evaluation panel and environmental risk evaluation panel. This third panel could be used at first to evaluate every single gmo organism, but later on it can do this at an aggregated level of groups of similar genetically modified organisms.

The Health Council will further explore the option of a broad agricultural biotechnology assessment.

References

(1) (CA03) Codex Alimentarius. Report of the third session of the Codex ad hoc intergovernmental task force on foods derived from biotechnology. Alinorm 03/343. Codex Alimentarius Commission, Geneva 2003.

(2) (Dip99) ILSI: Diplock A.T. et al. Scientific Concepts of functional food in Europe: consensus document. B J Nutr 1999; 81: suppl 1

(3) (EG00) White Paper on Food Safety. COM (1999) 719 def., 12 January 2000

(4) (EG01) Directive 2001/18/EC of the European Parliament and of the Council of 12 March 2001 on the deliberate release into the environment of genetically modified organisms and repealing Council Directive 90/220/EEC. Official Journal of the European Communities 2001; L106: 1-38.

(5) (EG97) Regulation (EC) no 258/97 of the European Parliament and of the Council of 27 January 1997 concerning novel foods and novel food ingredients. Official Journal of the European Communities 1997; L43: 1-6.

(6) (EG97a) Commission Recommendation No. 97/618/EC of 29 July 1997 concerning the scientific aspects and the presentation of information necessary to support applications for the placing on the market of novel foods and novel food ingredients and the preparation of initial assessment reports under Regulation (EC) No. 258/97 of the European Parliament and of the Council; L253:1-36

(7) (Env01) Consensus document on compositional considerations for new varieties of maize (Zea mays): key food and feed nutrients, anti-nutrients and secondary plant metabolites. ENV/JM/Food (2001)1 November 1, 2001 (draft)

(8) (FAO96) Biotechnology and Food Safety. Report of a joint FAO/WHO Consultation. Rome, FAO 1996.

(9) (FAO01) Evaluation of allergenicity of genetically modified foods. Report of a joint FAO/WHO expert consultation on allergenicity of foods derived from biotechnology. Rome, FAO 2001.

(10) (GR92) Toxicological aspects of biotechnologically prepared products Committee. Product safety with new biotechnology. The Hague, Health Council 1992, publication number 1992/03.

(11) (Hel95) Hellenas KE *et al.* High levels of glycoalkaloids in the established Swedish potato variety "Magnum Bonum". J Sci Food Agric 1995; 23: 520-23

(12) (Kui01) Kuiper HA, Kleter GA, Noteborn HPJM, Kok EJ. Assessment of the food safety issues related to genetically modified foods. The Plant Journal 2001; 27(6): 503-28.

(13) (Löw94) Löwik MRH, Brussaard JH, Hulshof KFAM, Kistemaker C, Schaafsma G, Ockhuizen Th, Hermus RJJ. Adequacy of the diet in the Netherlands in 1987-1988 (Dutch Nutrition Surveillance System). Int J Food Sci Nutr 1994; 45(Suppl. 1): S1-S62.

(14) (Löw97) Löwik MRH, Hulshof KFAM, Heyden LJM van der, Brussaard JH, Burema J, Kistemaker C, Vries PJF de. Changes in the diet in the Netherlands: 1987-1988 to 1992 (Dutch Nutrition Surveillance System), Int J Food Sci Nutr 1997.

(15) (Meij97) Meijer GW, Westrate JA. Interesterification of fats in margarine; effect on blood lipids, blood enzymes, and hemostasis parameters. Eur J   Clin Nutr 1997; 51:527-534

(16) (NEVO01) Netherlands Nutrient Databank 2001. NEVO Foundation, Netherlands Nutrition Centre. The Hague, 2001.

(17) (OECD93) Safety evaluation of foods derived by modern biotechnology. Concepts and principles. Paris, OECD 1993.

(18) (OECD96) OECD Workshop on Food Safety Evaluation. Paris, OECD 1996.

(19) (OECD98) Report of the OECD workshop on the toxicological and nutritional testing of novel foods. Paris, OECD 1998.

(20) (OECD98a) Guideline for the testing of chemicals. Repeated dose 90-day oral toxicity study in rodents. Paris, OECD 1998.

(21) (OECD00) Report of the task force for the safety of novel foods and feeds. Paris, OECD 2000.

(22) (SCF99) Opinion concerning the scientific basis for determining whether food products, derived from genetically modified maize, could be included in a list of food products which do not require labelling because they do not contain (detectable) traces of DNA or protein. Brussels, Scientific Committee on Food of the EU 1999

(23) (SCF00) Opinion of the Scientific Committee on Food on a request for the safety assessment of the use of phytosterol esters in yellow fat spreads. Brussels, Scientific Committee on Food of the EU, 2000.

(24) (SSC99) Opinion of the Scientific Steering Committee on microbial resistance, Brussels, Scientific Steering Committee of the EU 1999.

(25) (Tam00) Tammi A, Rönnemaa T, Gylling H, Rask-Nissilä L, Viikari J, Tuominen J, Pulkki K, Simell O. Plant stanol ester margarine lowers serum total and low-density lipoprotein cholesterol concentrations of healthy children: the STRIP project. J Pediatr 2000; 136: 503-510.

(26) (Tay01) Taylor SL, Hefle SL. Will genetically modified foods be allergenic? J Allergy Clin Immunol. 2001 May;107(5):765-71. Review.

(27) (TK00)   Beleidsnota Biotechnologie. Ministers van VROM, EZ, VWS, LNV en OCW.. Den Haag, Tweede Kamer der Staten Generaal 2000; 27428 nr 1.

(28) (Tru90) Trumble JT.et al. Host plant resistance and linear furanocoumarine content of Apium accessions. J Econ Entomol 1990; 83: 519-25

(29) (USDA02) Continuing surveys of food intakes by individuals, 89-92CSFII. United States Department of Agriculture

(30) (VR93) Commissie Biotechnologie. Advies inzake Biotechnologie. Den Haag, Voedingsraad 1993, publicatie nummer 707

(31) (WHO91) Strategies for assessing the safety of foods produced by biotechnology. Report of a joint FAO/WHO Consultation. Geneva, WHO 1991.

(32) (WHO00) Safety aspects of genetically modified foods of plant origin. Report of a joint FAO/WHO expert consultation on foods derived from biotechnology. Geneva, WHO 2000.

SESSION 3.2:
STATE OF THE ART IN OTHER COUNTRIES

AGRICULTURAL GMOs: RISK ANALYSIS AND INTELLECTUAL PROPERTY PROTECTION IN THE USA.

Alan McHughen
FACN. University of California
Riverside, Ca 92521, USA.

1. Introduction

The United States has a long history of encouraging innovations requiring intellectual property protection and also of analyzing and managing risks associated with those innovations. Although the history of genetic engineering is relatively recent — just over 30 years—the approach taken by the US is to adapt existing legal and regulatory frameworks to encompass biotechnological innovations.

Most other jurisdictions, in contrast, decided to develop new regulatory structures exclusively for biotechnological innovations. As well, the approach of the US is to evaluate the risks associated with the tangible product of biotechnology, while most other nations attempt to evaluate the risks associated with the processes of biotechnology.

Although all share the same legitimate concerns, to ensure that biotechnology is safe both for humans and the environment, these differing approaches, when applied to commodities in international trade, create friction among trading partners due to their fundamentally different and incompatible foundations.

2. Problems with differing definitions

Premarket regulatory scrutiny is designed to identify and mitigate potentially hazardous products before they get onto the market. A maxim of regulatory theory is that the degree of regulatory scrutiny should be commensurate with the degree of risk. However, some products in some countries face enormous regulatory scrutiny, while in other countries the same product, conceivably carrying the same or similar degree of risk, receives little or no regulatory scrutiny.

For example, consider herbicide tolerant canola ("double-zero" rapeseed). Figure 1 lists canola with tolerance to one (or more) of six different herbicide groups. A canola cultivar with resistance to, for example, sulfonylurea (group 2 herbicide) presents concerns for food and feed safety. That is, humans and other animals face risks from ingesting sulfonylurea chemicals. Will the dose needed to provide effective weed control in the crop present unacceptable food and feed hazard? As well, there are legitimate environmental concerns. Will resistance to sulfonylurea give the canola cultivar increased fitness, or make it more aggressive, or more weedy, thus presenting a risk to escape and establish as an invasive weed? Will the HT canola cross pollinate with related weeds, making them even more noxious and hard to control?

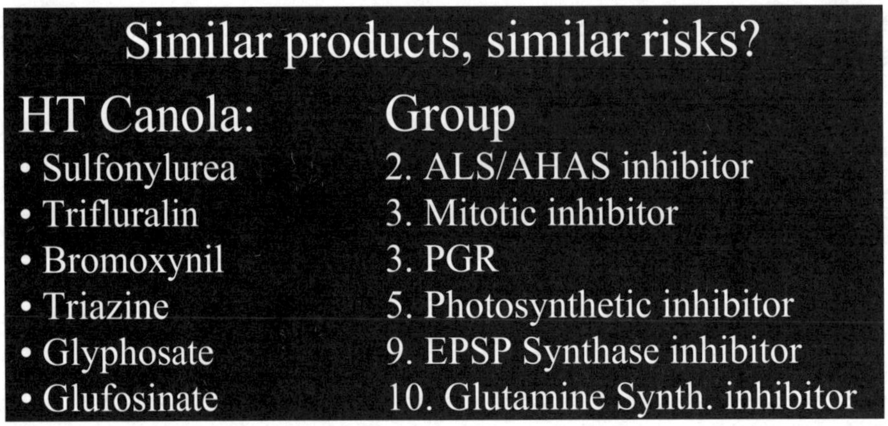

Similar products, similar risks?	
HT Canola:	**Group**
• Sulfonylurea	2. ALS/AHAS inhibitor
• Trifluralin	3. Mitotic inhibitor
• Bromoxynil	3. PGR
• Triazine	5. Photosynthetic inhibitor
• Glyphosate	9. EPSP Synthase inhibitor
• Glufosinate	10. Glutamine Synth. inhibitor

Figure 1. Canola (rapeseed) cultivars are available with tolerance to six families of herbicide. Some faced greater regulatory scrutiny than others, but not necessarily based on the risk posed by the herbicide or by the tolerance mechanism.

These health and environmental safety questions and concerns are legitimate and must be addressed by competent regulatory authorities. The other examples, using different herbicides, evoke similar legitimate questions and concerns. One can argue the relative degree of risk between, say, the bromoxynil tolerant canola and the glyphosate tolerant canola, and regulatory agencies do just that. Some chemicals are less hazardous, either to humans or the environment, than others.

Each of these examples presents differing degrees of risk and may raise different issues, which should be evaluated by competent regulatory authorities prior to commercial release. But, depending on jurisdiction and foundation of regulatory oversight, a herbicide tolerant canola cultivar can come under stringent regulatory control in one place, but the same canola cultivar could receive minimal regulatory scrutiny in another.

The same product, presenting the same risks, is regulated differently in different jurisdictions. If we agree that regulatory scrutiny should be commensurate with degree of risk, someone is in the wrong, either over-regulating or under-regulating. It may be tempting to invoke precaution, and support over-regulation to guard against under regulation. However, over-regulation is just as irresponsible and dangerous, because all regulatory agencies are constrained by finite financial and human resources, so over regulating in one arena necessarily means under regulating in another. More importantly, an inappropriate degree of regulation (either too much or too little) jeopardizes public credibility and trust, particularly when mistakes become public knowledge and the public comes to consider 'competent authority' as an oxymoron.

All jurisdictions are concerned about the risks associated with genetically modified organisms GMOs. However, the international community cannot agree on what is a GMO (or relevant term to capture and regulate the supposed high risk articles).

In the US, a GMO includes those plants developed using conventional breeding methods to change the genetic makeup. But regulated articles are those produced by recombinant DNA and carry modified DNA or protein.

Canada, in contrast, regulates based on novelty of the introduced characteristic. Plant with novel traits, PNTs, capture most, but not all, products of rDNA, along with some products of conventional breeding (http://www.inspection.gc.ca/english/plaveg/pbo/pntchae.shtml).

In parts of Africa, products of tissue culture, somaclonal variants, etc. are considered products of biotechnology.

In Europe under both old (90/220/EC) and new (2001/18/EC) regulatory directives a GMO *"means an organism... in which the genetic material has been altered in a way that does not occur naturally by mating and/or natural recombination...."*. This definition encompasses more than rDNA, including, for example, injection of genetic material and fusion of cells that would not be able to hybridize in nature. Unclear, however, is why the regulations exempt traditional interspecific grafts, chimeras, and irradiation-induced mutants, which provide many common foods and are clearly unnatural, but do capture foods from rDNA involving transfer only of gene material from the same species, such as a rice gene transferred into another rice variety, which could easily occur in nature.

With such divergent definitions triggering either intense regulatory scrutiny or little, products considered innocuous in one jurisdiction might invoke rigorous examination in another. Meanwhile, the actual threat, if any, posed by the product to health or environment remains constant.

All of the herbicide tolerant canola cultivar examples in Figure 1 are real; they're available and grown in Canada. Of these six, three are products of rDNA, one is a spontaneous mutant, one is natural and common across the genus, and one has been developed using mutation breeding. Each one presents different risks and should be assessed separately, because of the differing features, not because of the method of breeding.

3. Risk assessment, analysis, and management in the USA

Three primary agencies oversee the regulation of products of agricultural biotechnology in the USA (Table 1). The US Department of Agriculture (USDA), under the statutory authority of the federal Plant Pest Act assesses and manages the environmental risks of genetically engineered plants prior to commercial release or transborder transport.

Using the regulations detailed under 7 CFR 340 - *Introduction of Organisms and Products Altered or Produced Through Genetic Engineering Which are Plant Pests or Which There is Reason to Believe are Plant Pests*, USDA interprets "plant pest" to include any organism that might damage crops up to but short of human beings. It gives the USDA considerable power to regulate *"any ... organism or product altered or produced through genetic engineering which the Administrator determines is a plant pest or has reason to believe is a plant pest"*. The USDA regulatory review is mandatory for all genetically engineered plants. Since 1994, the USDA has approved approximately 55 GE genotypes for unconfined environmental release.

VARIETY RELEASE REQUIREMENTS: GENETICALLY ENGINEERED CROPS
USDA - *environmental issues*
<ul><li>Plant Pest Act (PPA); *also administers*</li><li>Plant Patent Act</li><li>Plant Variety Protection Act (PVPA)</li></ul>
FDA - *food and feed safety*
<ul><li>Federal Food, Drug, and Cosmetic Act (FFDCA)</li></ul>
EPA - *pesticide usage, food safety issues*
<ul><li>Federal Insecticide, Fungicide, and Rodenticide Act (FIFRA)</li><li>Federal Food, Drug, and Cosmetic Act (FFDCA) and</li><li>Toxic Substances Control Act (TSCA)</li></ul>

Table 1. Regulatory agencies in the US and the statutory authority for regulating genetically engineered crops.

The Food and Drug Administration (FDA) is responsible for the overall safety and security of the food supply. Their primary concern is the health and nutritional equivalency of new foods, including those developed using biotechnological methods. The presence in foods of adulterants, novel toxicants or allergens, or a substantial change in amounts of nutrients provides statutory authority, under the Federal Food, Drug, and Cosmetic Act (FFDCA) for FDA regulatory action. Regulatory review is based on the composition of the food (FDA, 1992). While the pre-market consultation between FDA and the developer of a new crop is legally 'voluntary', the consultation/review process is *de facto* mandatory, as all GE foods on the US market were evaluated by FDA. Nevertheless, a formal proposal to make the process mandatory is under consideration.

The Environmental Protection Agency (EPA) has authority under several statutes, including the Federal Insecticide, Fungicide, and Rodenticide Act (FIFRA), the Federal Food, Drug, and Cosmetic Act (FFDCA) and the Toxic Substances Control Act (TSCA) for regulating changes in chemical uses in the environment and in food. In this context, it conducts a mandatory review for products involving new or altered patterns of pesticide use. Most commercial GE crops express additional pesticide tolerance (e.g. introduced resistance to a particular herbicide) or are themselves pesticidal (e.g. GE Bt crops), so EPA invokes mandatory regulatory review for nearly all GE crops.

3.1. Product vs. Process

While the trigger for regulatory action in the US remains the process of rDNA, the analysis is of the product resulting from rDNA. That is, the US regulatory agencies conduct their analyses on the physical manifestation of rDNA technologies, arguing that the process by which the product was made is irrelevant. Put another way, in contrast to the EU, the US considers there is no inherent hazard with rDNA as a methodology, but the product resulting from the methods may pose an actionable risk.

The EU and some other jurisdictions regulate based on the risks associated with the process of rDNA. The operating assumption is that 'natural' breeding processes (including, astonishingly, ionizing radiation mutagenesis) are deemed safe and so do not warrant regulatory oversight, and that gene recombination, even when hybridizing genes that could readily hybridize naturally, requires stringent regulation. Numerous studies have shown that products of so-called natural breeding are not inherently or invariably safe.

The recent National Academy of Science (NAS) report on the environmental effects of transgenic plants (NAS, 2002) discusses several examples of environmental damage caused by 'natural' plant products, and Cheeke (1998) considers several examples of health threats from 'natural' breeding, including Lenape potato, unintentionally selected for excess glycoalkaloid content and withdrawn from commercial release (Zitnak and Johnston, 1970) but still used as a parent in potato breeding programs.

Other common examples includes celery with elevated psoralens, which caused rashes on farmworkers' skin, but nevertheless selected because of the pest fighting qualities (Diawara et al, 1993) and the heritage variety Magnum Bonum potato, which was fine until cool weather stimulated production of toxic levels of glycoalkaloids (Van Gelder et al., 1988; Hellenaes et al, 1995).

The dichotomy between the attitude that rDNA (= high risk) and natural (= low risk) seems unsupported by available evidence. Furthermore it seems a stretch of credibility that all diverse forms of rDNA are equally high risk, considering the fundamental differences between, say, the natural, biological DNA delivery system of *Agrobacterium* and the unnatural, physical DNA delivery system of the particle gun. The mechanisms of each process are different in almost every respect, other than providing a DNA fragment to a target host cell.

Equally puzzling is the assertion that two divergent products, clearly differing in hazard, triggers the same regulatory scrutiny simply because rDNA was the method of gene transfer. For example, a rice cultivar transformed using rDNA to deliver an additional rice gene is considered *a priori* equally hazardous to an rDNA rice variety carrying a novel toxicant.

A recent movement in the US and other countries is the recognition that all new organisms carry some degree of risk, and that that risk is based on the new features of the organism, the nature of the species carrying the new feature and the environment in which it is grown (NAS, 2002).

In following this line of reasoning, it is apparent that the method of derivation is irrelevant, so exclusive regulatory focus on GM or rDNA or traditional breeding is meaningless. A new variety of rice carrying the Xa21 gene for disease resistance presents the same risk whether it was developed using rDNA or ordinary cross breeding.

However, in most jurisdictions, including the US, the rDNA rice will receive intense regulatory scrutiny prior to commercial release, and the conventional will receive almost none. At the same time, a new transgenic rice variety carrying a gene from scorpion will receive the same regulatory scrutiny as a rDNA rice carrying only an additional rice Xa21 gene.

Does it make sense to load all products of one method of breeding with substantial regulatory burden when some products are demonstrably riskier than others? Does it make sense to exempt from regulatory scrutiny all products of certain methods, when it is known all such products carry environmental or health risks? For example, conventional breeding has resulted in food products hazardous to health (solanine in potatoes, psoralens in celery), and crop cultivars with devastating effects in the environment (NAS 2002). Yet regulations are designed only to capture products of rDNA methods.

Several authorities are suggesting ranking products of rDNA according to risk level, and applying regulatory scrutiny according to true degree of risk imposed. There are real environmental concerns based on the novelty of the phenotype, the host species, and the geography of release area in terms of invasiveness or reproductive success.

Hancock (2003) takes the initial discussion of relative environmental risks discussed in the NAS (2002) report further. He describes six levels of concern for invasiveness of plant species in North America, from those species with no compatible relatives and few weediness traits (e.g. cabbage, soybean) to those with many compatible wild relatives and highly competitive and invasive traits (e.g. canola, oats). He then considers five levels of fitness impact of new traits, from benign (e.g. marker genes) to potentially highly fitness enhanced (e.g. environmental stress tolerance). Then, by considering the combination of host plant species with the new trait, a defensible and sensible degree of regulatory scrutiny may be applied. Similarly, Strauss argues to consider regulating new crops, based on the real risks, from low confinement needs (such as same species gene transfers) to high (with novel toxic, allergenic or pharmaceutical substances) (Strauss, 2003).

3.2. Substantial Equivalence, Precautionary Approach

Recognition that some plants are more hazardous than others, regardless of method of breeding, brings us to the utility of using substantial equivalence with a precautionary approach as a basis for regulatory scrutiny. Contrary to popular opinion, substantial equivalence is not an endpoint, but provides a starting point for regulatory questions (Kuiper, 2003).

The OECD developed substantial equivalence several years ago as a means to approach a new product in attempting to determine the degree of risks imposed (OECD, 1993). In it, one considers the base or parental species and the hazards associated with it, in comparison with the new organism and its new features, and the hazards associated with them. In this manner, a transgenic soybean with a novel herbicide tolerance is compared with a traditional soybean plus the appropriate herbicide tolerance. Substantial equivalence provides a platform or baseline upon which to subsequently compare the new organism with the established version. Considering the new soybean in isolation denies the fact that the organism is still basically a known quantity, a soybean, but with an additional feature. Without the baseline, one might be forced to evaluate the new soybean is if it were an entirely new organism from outer space.

3.3. Labeling of Food Products

Mandatory labels in the US are based on the physical, chemical or biological properties of the food as consumed. The required label lists nutritional information, plus any toxic or allergenic constituents. This requirement is applied to all packaged foods, including those derived from rDNA technologies. However, the process of breeding or manufacture, unless it leads to changes in physical, chemical or biological characteristics in the food, is not subject to mandatory labeling. To meet consumer demand for non-safety related information, voluntary labels serve well. Organic, kosher, halal are all effective, yet voluntary, labels on various US foodstuffs.

US regulators enjoy remarkable public support and trust. One of the reasons the debate over biotech is muted is that American consumers have not had to contend with outbreaks of Mad Cow disease, *Salmonella* in dairy products, dioxins in feeds, foot and mouth disease or other true public health threats seen in Europe in recent years.

Because of the excellent track record, the average American is content to trust the USDA, FDA and EPA with making the regulatory decisions. In turn the US regulators, wishing to maintain their excellent track record and high degree of public confidence, are likely to continue regulating on the basis of demonstrable hazards, and expend less energy on publicly perceived risks for which there are few scientifically valid concerns.

4. Intellectual Property Rights

Intellectual property rights in the US provide multiple and combinable means of protecting new living organisms, including plant genotypes and cultivars.

Major statutes include the UPOV convention-compliant Plant Variety Protection Act, the Plant Patent Act and the utility Patent Act (the traditional means to protect inventions). These have differing applications and criteria, but a breeder can enjoy protection from both a utility patent as well as a plant patent if the respective criteria are satisfied.

The Plant Variety Protection Act (PVPA) enacted in December of 1970, was amended in 1994 to conform to the provisions of UPOV. This Act, administered by USDA, offers certificates of protection covering new tubers and seed reproduced plant varieties. Criteria include DUS: the new variety must be Distinct from previous varieties, Uniform in a population stand and genetically Stable over successive generations.

The certificate is valid for 20 years from time of application for most crop species, and 25 years for trees, shrubs and vines. Under this form of protection, the certificate holder (usually the breeder or breeder's employer) can prohibit others from selling, marketing, importing, exporting, offering, delivering, consigning, exchanging, or even possessing the variety without explicit permission.

PVPA carries a couple of important exemptions, neither of which applies to either utility patents or Plant Patents:

1. Research Exemption. Other breeders or breeding institutions can use the protected variety in research programs, particularly to derive new varieties, which ultimately may themselves be protected under PVPA.

2. Farmer's Exemption. Farmers (and home gardeners, for example) can save seed from one harvest to replant on the same farm in subsequent years. However, unlike in some jurisdictions, the farmer's exemption does not permit the farmer to sell that seed.

The Plant Patent Act (PPA) of 1930, amended in 1995, protects asexually reproduced plant varieties developed using breeding techniques or identified in a cultivated field (but not a novel plant discovered or 'found' in the wild).

Plants may include those developed and vegetatively propagated from sports, mutants or hybrids, as well as those discovered or found, as long as the new plant is so discovered or found in a cultivated area.

Like varieties protected under PVPA certificates, the plant patent requires DUS– the new plant must be Distinct, Uniform and Stable over propagation generations. Again, with seed varieties protected under PVPA, the period of protection is 20 years from date of filing. The plant patent gives exclusivity to the holder and prohibits others from propagating and selling the plants and plant parts, such as buds, scions, etc.

Utility patents. Recognition for protection of inventions dates back to the US Constitution (US Patent Act United States Patent Act (28 U.S.C. §§1295, 1338). As in most nations, utility patents were the domain of traditional inventions, products of innovative engineering and manufacture.

This changed in recent years, when living things, which previously were considered "non-patentable subject matter" became eligible for utility patent protection. Like most other nations, utility patents could be awarded for an invention if it fulfilled the three basic Trips (1996) criteria: novelty, utility and non-obviousness (equivalent to the 'inventive step' of other patent language).

The Chakrabarty decision of 1980 (Diamond v. Chakrabarty) opened the doors of the patent office to living things, at least microbial living things. The patentability of higher organisms was approved in 1985 with an elevated tryptophan producing line of maize (see Hibberd, 1985) and finally extended even to non-human mammals in 1988 with the successful patenting of the so-called Harvard mouse, or oncomouse (Leder and Stewart, 1988).

Another method of intellectual property (I.P.) protection is the contract. A contractual agreement is a higher order of protection, as signatories may sign away at least some rights they otherwise might enjoy. In our context, the Technology Use Agreement (TUA) is a contract between Monsanto and farmers wishing to grow Monsanto's seeds. In it, the farmer agrees to provisions granting Monsanto considerable control over the farmer's practice. In return, the farmer gets to grow the desired seed. The farmer need not sign the contract, but in that case will not legally be able to grow the seed.

An additional mechanism of intellectual property I.P. protection in the seed business is the use of hybrid seeds. Common in maize for over 50 years, hybrids are high performing cultivars.

Farmers buy and grow hybrid seeds, and typically harvest greater yields than if they had grown non hybrid (called "open pollinated") varieties. The protection is built into the seeds. Because of the nature of the hybrids, a farmer who invokes his traditional right to save and plant seed from a hybrid plant is disappointed in the poor performance and the variability of the resulting crop.

The farmer wishing to continue successful crops must purchase fresh seed each year, as the hybrid characteristics last only one generation. The companies generate new hybrid seed at special farms using special parental lines, the exact genetic background of which is maintained in high secrecy.

In any case, a farmer growing such hybrids is legally entitled to replant seed from the hybrid crop, but is usually so disappointed in the result that he or she returns to buy fresh hybrid seed for the following season.

In addition to these standards forms of I.P. protection, some firms also employ trade secrets as a form of I.P. protection to circumvent the public disclosure and public domain deposition provisions of utility patents.

The exact recipe for Coke has been maintained as a trade secret for years, giving Coca-Cola™ a virtual and continuing monopoly since the beginning of the company, but a utility patent would have given them exclusivity for only 17 years, long expired by now.

Trade secrets might be suitable for recipes and processes, but are unlikely to help preserve a self replicating invention like a seed. However, parent lines of f1 hybrids, since they themselves are not marketed under either UPOV or patent protection, need not be exposed to public disclosure or access by others. Their exact genetic identities are instead protected by what is essentially a trade secret.

An unusual provision of US law is the ability to protect I.P. using under more than one means. A plant variety can conceivably be protected both under utility patent and a PVPA certificate (Pioneer et al, 2000).

In Europe, plant varieties can be protected as a registered variety, but cannot themselves be patented. Genotypes, non commercial strains, may be patented, however, and then cultivars derived from the genotype may be protected as varieties.

One of the concerns in the public debate over biotechnology is that companies might go into traditional farming areas and patent longstanding varieties of landraces, then force local farmers to give up their traditional practice of saving seed, instead making them purchase seeds – the same seeds they'd grown for generations – from the company. A horrifying thought, and fortunately not possible under patent laws.

For one thing, a US patent is in effect only in the USA. A company patenting an invention in the US, but not elsewhere, has no case against someone using the invention outside of the boundaries of the USA. A seed variety patented but not protected under plant breeders' rights may be legally grown elsewhere without infringing on the US patent.

Also, and more important, the first criterion in the utility patent list is novelty. A patent cannot be granted on an item already in common use. A patent may be granted on a new, improved version of the cultivar, but the current variety is not patentable.

This is the basis of a recent controversial US patent (# 5, 663,484) on basmati rice issued to RiceTec, Inc (Sarreal, et al., 1997). Unlike popular concern, this patent did not cover traditional basmati rice, but rather the allowed claims included three new improved cultivars developed by the company. Unfortunately, widespread misunderstanding of patent law and practice lead to considerable unnecessary and continuing public anxiety.

Intellectual property rights and regulation of risks is perfect nowhere. However, all US agencies are under near-constant review to improve efficacy and safety, encourage innovation and instill public trust. Increasing familiarity with risks from products of biotechnology relative to those of conventional technologies, and adopting means of appropriate protection for innovations will ensure continued public support and confidence for the future.

References

(1) Canadian Food Inspection agency (CFIA). Regulation of plants with novel traits. http://www.inspection.gc.ca/english/plaveg/pbo/pntchae.shtml

(2) Cheeke, P. 1998. Natural Toxicants in Feeds, Forages, and Poisonous Plants. Interstate Publishers Inc, Danville, IL USA.:

(3) *Diamond v Chakrabarty* [447 U.S. 303, 206 USPQ 193 (1980)].

(4) Diawara, M.M.; Trumble, J.T., and C.F Quiros. 1993. Linear furanocoumarins of three celery breeding lines: implications for integrated pest management. Journal Agric and Food Chem. 41 (5): 819-824.

(5) FDA (Food and Drug Administration) (USA). 1992. Statement of policy: foods derived from new plant varieties. Fed. Register 57:22984.

(6) Hancock, J.F. 2003. A framework for assessing the risk of transgenic crops. BioScience 53:512-519.

(7) Hellenaes, K.E., C. Branzell, H. Johnsson and P. Slanina. 1995. High levels of glycoalkaloids in the established Swedish potato variety Magnum Bonum. Journal of Science of Food and Agriculture 68: 249-255.

(8) Hibberd, K. *et al.*, 1985. Patent and Trademark Office Board of Patent Appeals and Interferences. Opinion dated September 18, 1985 as corrected September 24, 1985

(9) Kuiper, H. 2003. The use of profiling methods for identification and assessment of unintended effects in genetically modified foods. National Academy of Sciences Workshop on identifying unintended health effects of genetically engineered foods, February 6-8, 2003. Washington, DC.

(10) Leder, P. and T. Stewart, 1988. A transgenic non-human eukaryotic animal whose germ cells and somatic cells contain an activated oncogene sequence introduced into the animal, or an ancestor of the animal, at an embryonic stage. The so-called Harvard mouse patent, USP 4,736,866

(11) National Academy of Sciences, 2002. Environmental effects of transgenic plants: the scope and adequacy of regulation. National Academy Press. Washington, DC.

(12) Organization for Economic Cooperation and Development (OECD), 1993. Safety evaluation of foods derived by modern biotechnology. Paris, 1993

(13) Pioneer Hi-Bred International Inc. v. J.E.M. AG Supply Inc., 2000. US Patents Quarterly 2nd Edition, vol. 53, p 1440. (Fed. Cir. 2000). Available at the Federal Circuit Web site: http://www.fedcir.gov/

(14) Sarreal, E. S., J. A Mann, J.E. Stroike, and R. D. Andrews. 1997. Basmati rice lines and grains. USPTO patent 5,663,484. September 2, 1997.

(15) Strauss S.H. 2003. Genetic technologies: Genomics, Genetic Engineering, and Domestication of Crops. Science 300: 61-62

(16) TRIPS (Trade related aspects of international trade) 1996. World Intellectual Property Office, Pub. No. 223 (E).

(17) US Patent Act United States Patent Act (28 U.S.C. §§1295, 1338)

(18) US Plant Variety Protection Act: http://www.ams.usda.gov/science/PVPO/pvp.htm and http://www.ams.usda.gov/science/PVPO/PVPO_Act/whole.pdf

(19) Van Gelder, W.M.J., J.H. Vinke and J.J.C. Scheffer, 1988. Steroidal glycoalkaloids in tubers and leaves of *Solanum* species used in potato breeding. Euphytica 38:147-158.

(20) Zitnack, A. and G.R. Johnston, 1970. Glycoalkaloid content of B5141-6 potatoes. American Potato Journal, 47:256-260.

JAPANESE VIEWS ON RISKS OF BIOTECHNOLOGY AND INTELLECTUAL PROPERTY

Masakazu Inaba (1) and Darryl Macer (2)
(1) Doctoral Program in Life and Environmental Sciences, University of Tsukuba,
1-1-1   Tennoudai, Tsukuba Science City, Ibaraki, 305-8572, Japan.
(2) Eubios Ethics Institute, Tsukuba Science City, 305-8691, Japan

Abstract

One of the central questions in the development of international biotechnology policy is whether persons have the same concepts of benefit and risk between countries, and how these views influence the policy. The first part of this paper presents analysis of comments on the benefits, risks and moral acceptability of biotechnology in Japan as viewed from different sectors of society. Two samples were obtained from mail response surveys in the year 2000, from the general public (N=297) and scientists (N=370), and one from the general public in 2003 (N=377). Comparison was made for a series of four questions on utility, risk, moral acceptability and overall encouragement, for applications of technology. The questions requested both agreement with a 5-point self-indicated scale, and the reasons behind these attitudes through open comments. The most acceptable of the applications were medicines produced in genetically modified microorganisms and a transgenic cancer mouse for research use. These were perceived to bring benefits by two thirds of the public in 2000 and 85% of the scientists. The least acceptable application was xenotransplantation, which even many of the scientists considered to be unnatural. Embryonic genetic diagnosis was more acceptable than xenotransplantation, however, more respondents saw ethical concerns with this application.

The second part of the paper will review some of the policies and practice for intellectual property protection from biotechnology in Japan. These include statements from bodies like the Japan Bioindustry Association (JBA) on benefit sharing, and positions that have been taken by scientists in debates over gene patenting. Japanese companies are among the world's leading holders of patents on human genes, but this is not necessarily a reflection of the views of most academics on whether products of biotechnology should be patented.

Finally, the bioethical maturity of different sectors of the Japanese society, public, scientists, policy makers, can not be measured by merely education level. It needs to be investigated in the context of the culture of information and high degree of distrust in the system that surveys reveal. Scientists tended to mention ethical issues less than the general public, which raises questions over whether scientists should really represent the views of the "public" in policy making in these areas, as they do in Japan. There is no real evidence that the concept of risk is dealt with differently in Japan by individual persons, however, some policy practices suggest Japanese companies are extremely cautious over the use of genetic engineering in products. Some comments from persons interviewed in different sectors of the community shed light on this.

1. Introduction

Japan has a population of 125 million persons enjoying a relatively high standard of living internationally, being the eighth most populated nation globally. Accordingly on the FAO index of food intake Japan rates as a developed country. In 1995 there was 4.282 million ha of land under crops, so the ratio of agricultural land per person is only 0.3 hectare per person, because the country is 80% mountainous. Of these crops 2 million hectare is under rice. In order to feed these people most food is imported.

Although some surveys on Japanese biotechnology have pointed out the relatively low importance of agricultural biotechnology when compared to agricultural exporting countries in Australasia or the United States, the increased capacity for food production from a limited area of land is of great potential benefit to Japan, where there is little agricultural land available. The government and industry have been promoting biotechnology since the 1980s.

The 2002 budget related to biotechnology in Japan included 27 billion Yen from the Ministry of Economy, Trade and Industry (METI), 128 billion yen from the Ministry of Health, Labor, and Welfare, 23 billion yen from the Ministry of Agriculture, Forestry and Fisheries, 71 billion yen from the Ministry of Education, Culture, Sports, Science and Technology (MEXT), and 4 billion yen from the Ministry of Environment (Japan Bioindustry Association Figures, 2002).

Given the large amount spent upon biotechnology in Japan, we can ask why almost nothing is spent discussing the ethical, social and legal (ELSI) issues raised by the application of biotechnology in society. Until now, there has been little spent on these issues when compared to other countries. For example in Canada 12% of the budget for the human genome project was spent on ELSI issues, and in USA 5%, but Japan has never got above 1% despite this point having been discussed internationally (Macer, 1992b).

The first part of this paper attempts to examine Japanese views of risk from random anonymous mail response surveys conducted over Japan in 2000 (Ng et al., 2000) and 2003 (Inaba and Macer, 2003). The results of some of these questions are presented, including analysis of results from a series of questions examining attitudes to six different applications of biotechnology, which articulates some of risk concerns people have. This type of analysis is useful for policy makers to know what are the real concerns of people, and help them to develop policies based on the facts. This method focuses on descriptive bioethics, that is, how do people think about biotechnology (Macer, 1996).

In addition, present Japanese policy on intellectual property, and how it was formed is discussed. This part focuses on the structure of policy making with the involvement of non-governmental organizations, especially the Japan Bioindustry Association (JBA). In addition there is discussion of the intellectual property strategic program of the Japanese government launched in July 2003.

2. Positive Support for Science and Technology in Japan

The results of public surveys are one method of descriptive bioethics. The national random samples across Japan in 1991, 1993, 2000 and 2003 have been described elsewhere (Macer, 1992b; Macer, 1994; Ng et al., 2000, Inaba and Macer, 2003). Sampling was done across all prefectures of Japan by using random mail sampling methods. In 1997 the random telephone survey method was used (Macer et al. 1997). The numbers of respondents were 297 public and 370 scientists in 2000, and 378 public in 2003. There is a mix of different sectors of the Japanese public, education, different occupations (not shown), and rural and urban populations in each sample. We estimate sample error at +/- 5%.

Table 1 shows the responses of the public to a general question on views of science and technology. There were few pessimists about science in general. Most people were concentrated in "more good" or "same", and people who thought science does more harm has been a low number in past decade. Besides, the majority thought that science makes and important contribution to the quality of life (90% in 1993, and 87% in 2000). Therefore, the general images of science and technology Japanese people had were positive. However, when people were asked specific questions, their concerns were revealed, and respondents differentiated several applications of biotechnology.

When people were asked whether most problems could be solved by applying more and better technology, less people expressed their disagreements (21% in 1993, 9% in 2000). However in 2000, 49% of the respondents answered "neither", and it became the majority (33% in 1993). Similar distribution was observed when people were asked a question regarding the utility of genetically modified foods against hunger. 44% answered "neither", and 37% "agreed".

Table 1: General pessimism about science remains low

Q3. Overall do you think science and technology do more harm than good, more good than harm, or about the same of each?

%	1990*	1991	1993	2003
N	2239	530	352	378
More harm	7	6	5	6
More good	53	55	42	43
Same	31	39	45	45
Don't know	10	-	8	7

*1990 (PMO survey data); 1991, 1993 and 2003 public surveys.

Table 2: Perception of the benefits and risks of genetic engineering 1993-2003

Q6. Do you personally believe each of these scientific discoveries and developments is a worthwhile area for scientific research? Why?... Y=Yes N= No DK=Don't know
Q7. Do you have any worries about the impact of research or its applications of these scientific discoveries and developments? How much? Why?...
W0=No W1= few W2=Some W3=A lot

%	Q6.Worthwhile area?			Q7.Worried about impact			
	Yes	No	DK	W0	W1	W2	W3
Computers							
2003	82	4	14	34	50	11	4
1993	85	3	12	57	34	7	2
In vitro fertilization (IVF)							
2003	56	18	26	15	48	25	12
1993	47	23	30	13	45	28	14

1991	58	21	21	21	29	23	18
Genetic Engineering							
2003	60	8	32	13	45	31	11
1993	57	10	33	22	39	24	15
1991	76	7	17	19	29	21	20
Pesticides							
2003	75	10	15	15	42	29	14
1993	84	9	7	21	36	26	17
1991	89	4	7	27	23	25	18
Cloning							
2003	28	27	45	7	24	27	41

On the other hand, when it comes to respondents themselves, 50% disagreed with eating foods containing GM ingredients in restaurants, and 32% stated "neither". The distribution shifted towards the negative side and the majority was not willing to eat GM foods. However, this question includes issues of open information such as labelling of food, and education that means understanding of the information, or consumers choice. Even then, the priority of food issue for respondents differs depending on circumstances.

Table 2 shows the results when people were asked about the benefits and risks of genetic engineering. Although many of respondents answered these areas were worthwhile, the level of their concerns increased slightly between 1993 and 2003 (Macer, 1994, Inaba and Macer, 2003). Table 3 shows the results with comparisons in attitudes over time between selected examples of GMOs for environmental release.

3. Perceptions of Risk Vary between Application of Biotechnology

Comparisons were made for a series of four questions on utility, risk, moral acceptability and overall encouragement, for each application. Respondents answered each question on six applications with open comments why they agreed, or disagreed. Out of six applications of biotechnology, two were agricultural applications, and the rest of four were medical applications:

1. modern biotechnology in the production of foods and drinks (biotech food),

2. gene insertion into a crop from other species to obtain pest resistance to make it more resistant to insect pests (pest-resistant crops)

3. bacteria inserted human genes to produce medicines (GM medicines)

4. transgenic mouse with cancer genes for research use (transgenic cancer mouse)

5. transgenic pig with human heart for xenotransplantation (xenotransplantation)

6. genetic testing using embryos for diseases such as cystic fibrosis (genetic testing).

In order to investigate risks from peoples' view, the balancing of risks and benefits and what kind of risks people perceive, were sought from analysis of the comments of respondents. In the survey's questionnaires, open spaces were attached below 5 point self evaluation scales, such as agree or disagree, in order to obtain the reason of their attitudes for each question. Analysis of open comments enables us to infer to what extent people will accept or will not accept new technologies considering the risks based on respondents' values. Because biotechnology sometimes casts questions relating to individual life values, analysis from socio-psychological based, or cultural/ individual value based view may be useful to understand what are perceived as risks.

Figure 1 shows the results of the surveys regarding awareness of biotechnology. The public awareness has increased compared with early 1990s although it has to be considered that the methods used for each survey were slightly different. A survey conducted over OECD countries shows that the level of Japanese interest in science and technologies that directly affect their daily lives was not as much as other OECD countries (Miller, 1996). However, the percentage of Japanese who are familiar with the term "biotechnology" may be highest among most countries (Macer et al., 1997). Overall, it is difficult to compare knowledge and perceptions between countries using different languages and styles. Other measures include sales of science magazines, use of science sites on the Internet, and the level of science reporting in the media.

Table 3: Concerns over release of GMOs
Q. "If there was no direct risk to humans and only very remote risks to the environment, would you approve or disapprove the environmental use of genetically engineered organisms designed to produce...?" (P: public, S: scientists)

%	P91	P93	P2000	P2003	S91	S2000
Tomatoes with better taste						
Yes	-	69	58.2	63	-	59.0
No	-	20	31.8	24	-	32.5
DK	-	11	10	14	-	8.5
Healthier meat (e.g. less fat)						
Yes	-	57	51.6	52	-	56.5
No	-	26	33.0	29	-	33.5
DK	-	17	15.4	18	-	9.9
Larger sport fish						
Yes	19	22	19.4	17	16.1	19.3
No	50	54	64	60	56.9	66.5
DK	31	24	16.5	23	27.0	14.2
Bacteria to clean up oils spills						
Yes	75	71	65.4	67	83.1	65.9
No	7	13	20.7	15	6.7	23.9
DK	18	16	13.9	19	10.2	10.2
Disease Resistant Crops						
Yes	75	66	54.5	51	85.7	60.7
No	6	17	28.7	24	5.0	26.1
DK	19	17	16.8	24	9.3	13.2
Cows which produce more milk						
Yes	-	44	42.1	37	-	59.7
No	-	32	39.6	35	-	29.0
DK	-	24	18.3	28	-	11.4

Figure 1: Public awareness of biotech is increasing

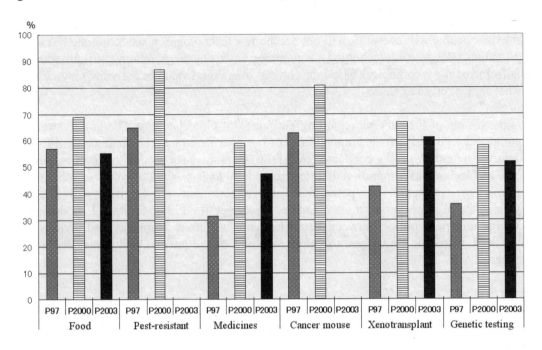

Figure 2: Trends of Japanese attitudes towards selected biotech applications

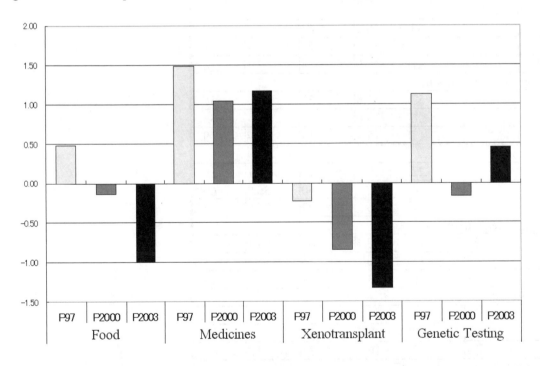

One question is how we can see the balance between perceived risk and perceived benefit. A calculation was made from the 5 point self evaluation scales in order to make a support index since it is useful to grasp the trends on how people perceive new technologies.

The application support index can be formulated as the sum of benefit, risk and overall encouragement of each application. Benefits and risks were often concrete and specific and respondents were expected to evaluate those significances. These two values of benefits and risks are risk/ benefit variables. However, attitudes of pros. and cons. towards each application can be formed besides the two variables. When people decide whether applications should be encouraged or not, other factors are often taken into consideration, such as ethical, moral, social, or cultural factors. Even if people express the same degree of utility and risk, overall encouragement could be different due to values which each respondent puts an importance on. Therefore, the support index includes variables of overall encouragement, and varies between the maximum +6 point in the cases of definitely agree, and the minimum –6 point in the cases of definitely disagree.

Figure 2 shows a time series of support index of Japanese public samples from 1997 to 2003. The figure shows clear decline of the acceptance of modern biotech foods and xenotransplantation. The most acceptable applications were medicines produced in genetically modified micro-organisms and a transgenic cancer mouse for research use. These were perceived to bring benefits by two thirds of the public in 2000 and 85% of the scientists (not shown). The least acceptable application was xenotransplantation, which even many scientists considered to be unnatural. Two agricultural applications and embryonic genetic diagnosis were more acceptable than xenotransplantation. However, more respondents saw ethical concerns with embryonic genetic testing.

Figure 3: The summary of the reasons for utility

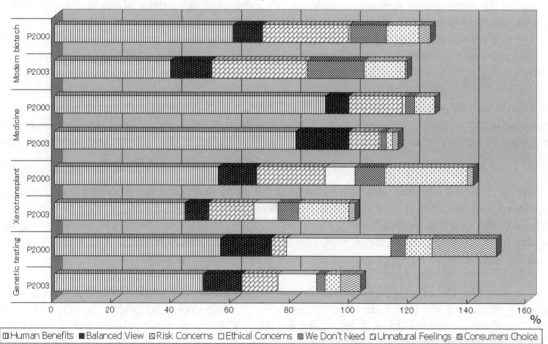

In order to investigate the reasons behind respondents' attitudes, open comments were analyzed by categorizing their comments into 7 broad categories. And comparison was made between the public results of 2000 and 2003. Figure 3 shows the results of utility of each application. One of the reasons why the acceptability of modern biotech food decreased is the decreased perceived utility.

Figure 4: The summary of the reasons for risk

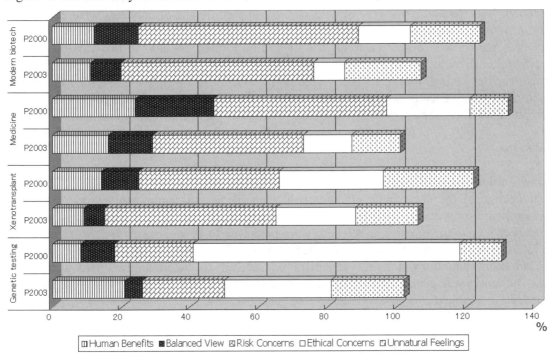

Figure 4 shows the summary of reasons for risk. In this case, perceived risks of modern biotech food have not been changed much. Therefore, the reason of decreased acceptability of the application partly was the decreased perceived utility, and not the increased perceived risks. On the other hand, the results of xenotransplantation shows that significantly increasing risks were the reason of decreased acceptability.

Although this is not shown here, the proportions of risk concerns through the questions were not significantly different between the general public and scientists in Japan. However, the degree of concerns would be different as we saw the differences of attitudes of the public and scientists from support index.

When people were asked whether applications of biotechnology were morally acceptable or not, they hardly expressed risk concerns. Rather emotional and ethical factors were more likely to be influential to their attitudes. The contrasts between risk concerns given in response to particular questions allows greater understanding of the concerns and hopes people have for biotechnology. The sum of human benefits and balanced view among applications are rough indicators of acceptability.

Overall the major risk concerns people raised in the open comments included: Environmental risks including other organisms and biosphere, physical risks which were mainly individual health, and social risks which were about possible impacts on human societies. Other concerns include fear of unknown, animal rights, insufficient check and control, or degradation of cultural and individual values. Some specifically mentioned the nature of long-term risk that raise greater doubts in their minds, and this can be applied to most of the specific risks. For example, a physical risk that was mentioned is a concern of genetic mutations in future generations.

This diversity of concerns was commonly seen in both public and scientists. As conclusions from survey, the descriptive bioethical approach reveals multidimensional concepts of risk from individual perceptions. There is wide diversity of concerns among every group surveyed. Different social and cultural perspectives may exist, but most reasons have no demographic predictor, and no significant gender, religious, age or educational predictors are observed. Therefore, we have few measures to predict concerns of people from their status at the moment, unless we actually ask them some key opinion questions as a classifier or examine their consumer patterns.

4. Japan Bioindustry Association and Promotion of Biotechnology by the Government

The second part of this paper discusses the strategy of intellectual property in Japan. The Japan Bioindustry Association (JBA) plays an important role in Japanese policy on science and technology. Considering the importance of JBA's advice, Prime Minister of Japan and His Cabinet launched the Strategic Program for the Creation, Protection and Exploitation of Intellectual Property in July 2003.

JBA is a non-profit organization and the main organization established though the support and co-operation of industry, academia, and government in Japan. JBA's predecessor body was the Japanese Association of Industrial Fermentation (JAIF), which was formed in 1942. This body provided places for exchanges between academic researchers and industrial technologists. In 1983, JAIF established a separate Bioindustry Development Center (BIDEC). In 1987, these two bodies emerged into JBA with the backup of METI. JBA promotes bioscience, biotechnology and bioindustry in both Japan and the rest of the world. Further it aims to promote the development and commercialization of new products and processes. JBA also functions as a think tank and platform for communication among scientists, technologists, policymakers and managers, and among academic, government and industry sectors.

JBA released a statement on benefit sharing in 2000, which contained three major points. The first point relates to the conservation of biological diversity, the second to the sustainable use of biological components, and the third point to the fair and equitable sharing of the benefits arising out of the utilization of genetic resources. The statement aims to develop the best practice for forming partnership between providers and users of genetic resources. The statement is written in the context of the Convention on Biological Diversity (CBD), especially relating to how developing countries protect their biological and genetic resources in the light of the TRIPS agreement of the WTO.

The statement articulates: 1. Recognizing the sovereign rights of States over their natural resources, the authority to determine access to genetic resources rests with the national government, 2. Access to genetic resources shall be subject to prior informed consent of the Contracting Party providing such resources, and 3. Access, where granted, shall be on mutually agreed terms. The concept of prior informed consent contends that joint research agreement is more desirable in the case where the provider of genetic resources has the intention and appropriate capacity for such research. The benefits, both monetary and non monetary ones (education, training and technology transfer), deriving from a joint research project should be shared by the provider of genetic resources and the user, according to the degree of contribution.

As for the pragmatic and effective procedures, the statement articulates that Each Contracting Party shall endeavor to create conditions to facilitate access to genetic resources for environmentally sound uses by other Contracting Parties and not to impose restrictions that run counter to the objectives of this Convention.

5. Other Government Policy to Promote Technology

Since the international competitive ability of Japanese industries decreased in the 1990s, the Japanese government has been reforming the structure of its economy, focusing on the importance of intellectual property, in order to recover the situation. There are few venture capital companies in Japan compared to the West. The number of Bioventures in Japan is 334 at the time of the year 2003 (Japan Bioindustry Association Report, 2003). MEXT launched the Science and Technology Basic Plan in 2001. This aims to create new knowledge for international contribution, to create vitality from the knowledge for sustainable development, and to create a fertile society with the knowledge for safe, secure and high quality life.

6. Intellectual Property Policy in Japan

The Prime Minister of Japan and His Cabinet set up intellectual property strategy council meetings and biotechnology strategy council meetings regarding GMOs in 2002. After the government received joint proposals from JBA and Japan Pharmaceutical Manufactures Association (JPMA), the Prime Minister of Japan and His Cabinet launched the Strategic Program for the Creation, Protection and Exploitation of Intellectual Property in July 2003.

This plan aims to realize a "nation built on intellectual property". The plan includes a purpose of stimulate the creation of intellectual property at universities, other academic institutions and industries by promoting the establishment of Technology Licensing Organizations (TLO) to reinforce technology transfer from academics to industries. TLOs are expected to take part in a business to manage intellectual properties, and to function to utilize inventions of universities through the co-operation between academics, industries, and the government. Other major aims are the establishment of the Law to Ensure Prompt Examination of Patent Application, which will help business to make quick judgments, and to establish the intellectual property high court to enhance the dispute settlement system. At the same time Alternative Dispute Resolution organisations will be reinforced in order to broaden options for Intellectual Property dispute settlements. Moreover, reinforcement of the intellectual property protection system is considered in the plan. This concerns medical related practices, inventions from Embryonic Stem cells or Embryonic Germ line cells as well as the protection of new plant varieties. The protection will cover new areas, such as novel creations, post genome inventions related to protein structure, etc. Some of these are still under discussion.

The Japanese government is reforming its patent system to be more harmonious with international standards because repeated applications are costly for inventors and the time framework for approval increases considerably. Overall, only 36% of total Japanese patent applications in life science of 2000 were also made abroad. This number is relatively higher than in other fields in Japan, such as industries (total 8.9% of patents application were made abroad between '94-'98) (JPO, 2002). However, 36% in life science is much less than for inventors in Europe (86%) and the US (90%) (JPO, 2003). The government continues to seek ways to stimulate more academics to apply for patents. Some of academics, however, are not favorable to obtain patents on genetic materials or genetic resources.

7. Reflections on Biotechnology Policy and Bioethics in Japan

The above discussion poses fundamental questions such as how we should deal with these issues and what kind of system is appropriate. Bioethical maturity exists in societies with many well informed and balanced persons. Bioethically mature means a person, or a society that can balance the benefits and risks of alternative options, make well-considered decisions, and talk about it. In all societies there is a transition from paternalism to consumer choice, and Japan is still finding its way through this transition. There is still a gap between the public and policy makers. There is a way to see the paradigm of social structure by considering what a mature society is. Japanese companies have been very cautious about introducing the products of biotechnology. In that sense more like Europe than USA. For example, some big companies stated that they don't use GM ingredients.

There is decreasing and low level of trust in almost anyone whether government or industry. The mail survey over Japan in 2003 revealed that more than 50% of the respondents answered "they don't trust their government or biotech companies" (Table 4). Even trust towards university professors and doctors have been decreasing in the past decade. This is a sign that the gap between the policy making side and the general public exists. The full potential of biotechnology has not been realized due to the lack of proper social mechanisms to resolve the issues into policy.

Table 4: Decreased trust comparing with the results of 1993 survey

Q26. Suppose that a number of groups made public statements about the benefits and risks of biotechnology products. Would you have a lot of trust, some trust, or no trust in statements made by...?
L: A lot of trust   S: Some trust   N: No trust

%	L1993	S1993	N1993	L2003	S2003	N2003
a. Government agencies	8	48	44	4	38	58
b. Consumer agencies	12	65	23	10	60	30
c. Companies making biotechnology products	6	43	51	5	34	61
d. Environmental groups	15	60	25	14	57	29
e. University professors	12	58	30	7	52	41
f. Medical doctors	12	58	30	7	54	39
g. UN organizations	-	-	-	17	57	26

On the other hand, there is some interesting comparative research on public attitudes towards risks. If compared to people in France and the US, Japanese were more likely to perceive the level of environmental risks, which are not controllable, higher than casual risks such as smoking and drinking, (Hirose, 1995). Tsuchida argues that Japanese perception of controllability over the environment is less than the counterparts because Japanese people think issues are beyond individuals, therefore, regulations exist to ensure the safety for a society (Tsuchida, 1997).

The idea of "environmental risk" was brought into Japanese policy in the early 1990s from the West and international institutions in the light of the Rio Declaration (Ikeda, 1997). The Japanese government enacted the Environment Basic Law in 1994 in order to protect the environment from wide use of chemicals. Discussion has occurred over the past few decades in different fora. However, at the time of writing (October 2003), the Food Safety Commission that was set up in July 2003 is still advertising publicly for more input on basic ideas for safety assessment on GMOs. It is organizing opinion exchange meetings on risk communication by inviting foreign experts. The framework of safety assessment in Japan is under discussion in the commission under existing legislation. Articles have been taken from Codex Alimentarius Commission, OECD, UK Food Standard Agency, World Health Organisation (WHO) and scientific journals. The Commission so far pointed out the gap of risk perception between experts and lay people, the importance of the role of mass media, how to redeem the trust to food policy of the government, and the need for guidelines for risk communication.

Since the USDA and U.S. Food and Drug Administration conducted risk studies in the 1980s and early 1990s there have been some shifts of the emphasis, with the increasing references to the precautionary principle in the EU. This may be due to public pressure, as well as cultural factors of an affluent society. The creation of a new NGO in July 2002 named "Life and Bio Plaza 21", a name similar to the 2000-

2001 focus group called the "Life and Bio Think Group" (Maekawa and Macer, 2001), was significant because it was focused on interacting with NGOs to seek a broader view than was possible under the JBA which has a strong industry links. Even if industry and government do not open their doors to the public fully, they have recognized a need to be more closely linked to NGOs. Another symbol of this closer link is the inclusion of consumer NGO representatives in Japanese government delegations to international bodies. For example in the 2002 meeting of the Codex Alimentarius ad hoc Task Force on Novel Foods Produced by Biotechnology.

The ad hoc Task Force on Novel Foods Produced by Biotechnology was set up for four years at the 23rd meeting of the Codex Alimentarius Commission, in June 1999. The Codex Alimentarius Commission is the joint FAO/WHO body that regulates the safety of food in the world. The task force successfully developed guidelines and standards for the safety assessment of foods produced using applications of biotechnology. The chair was held by Japan, considered as a point between the views of USA and the European Union (EU). Japan tries to be a bridge between the USA and EU on GM issues in other international fora as well, taking middle ground in some debates. Japan is less likely to break with world opinion than for example, the USA, which regards national autonomy as a higher ideal. The final guidelines (three) on risk assessment of foods produced by novel biotechnology were accepted by the full Codex Alimentarius Commission in 2003. Japanese people trust the UN much more than their own government, so perhaps the establishment of UN guidelines is an achievement of the Japanese government on risk assessment that will be more readily accepted than if they had regulated themselves simply on a national basis.

References

Note: references published by Macer are online at <http://www.biol.tsukuba.ac.jp/~macer/index.html>

(1) Chen Ng MA, Takeda C., Watanabe T. and Macer DRJ. (2000), "Attitudes of the Public and Scientists to Biotechnology in Japan at the Start of 2000", *Eubios Journal of Asian and International Bioethics* 106: 106-13.

(2) Ikeda, S. (1997), "Policy Science Toward Environmental Risk Management. Risk Regulation and Informed Choice", *The Society for Risk Analysis: Japan section* 8 (1): 81-88 (in Japanese).

(3) Inaba, M. & Macer, DRJ. (2003), "Attitudes to Biotechnology in Japan in 2003", *Eubios Journal of Asian and International Bioethics* 13(2003): 78-89.

(4) Inaba, M. and Macer, DRJ. (2003), "Japanese Views of Medical Biotechnology", pp. 178-196 in *Bioethics in Asia in the 21st Century*, Ed. Song SY., et al., Eubios Ethics Institute, Christchurch, New Zealand.

(5) Japan Bioindustry Association Figures (in the on-line Newsletterhttp://www.jba.or.jp/index_e.html)

(6) Japan Bioindustry Association. (2003), *The Report Regarding Bioventure and Biocluster* (in Japanese, title translated).

(7) Macer, DRJ. (1992a), *Attitudes to Genetic Engineering: Japanese and International Comparisons*, Eubios Ethics Institute, Christchurch, New Zealand.

(8) Macer, DRJ. (1992b), "The Far East of Biological Ethics", *Nature* 359, 770.

(9) Macer, DRJ. (1994), *Bioethics for the People by the People*, Eubios Ethics Institute, Christchurch, New Zealand.

(10) Macer, DRJ. (1996), "Public Acceptance and Risks of Biotechnology", pp. 227-245 in *Coping with Deliberate Release: The Limits of Risk Assesment,* Ed. A. van Dommelen, International Centre for Human and Public Affairs, Tilburg, The Netherlands.

(11) Macer, DRJ., Bezar, H., Harman, N., Kamada, H. and Macer, N. (1997), "Attitudes to Biotechnology in Japan and New Zealand in 1997, with International Comparisons", *Eubios Journal of Asian and International Bioethics* 7(1997), 137-151.

(12) Macer, DRJ. and Chen Ng, MA. (2000), "Changing Attitudes to Biotechnology in Japan", *Nature Biotechnology* 18: 945-7.

(13) Macer, DRJ., Azariah, J. & Srinives, P. (2000), "Attitudes to Biotechnology in Asia", *International Journal of Biotechnology* 2: 313-332.

(14) Macer, DRJ. (2002), "Patent or Perish? An Ethical Approach to Patenting Human Genes and Proteins", *The Pharmacogenomics Journal* 2: 361-6.

(15) Macer DRJ., Inaba, M., Maekawa, M., Ng, MC., and Obata, H. (2002), "Japanese Attitudes

Towards Xenotransplantation", *Public Understanding of Science* 11: 347-62.

(16) Maekawa, F., & Macer, D.R.J. (2001). Interactive Bioethics in a Focus Group on Life and Biotechnology in Japan. *Law and the Human Genome Review 15*, 173-201.

(17) Miller, John. (1996), "Public Understanding of Science and Technology in OECD Countries: a Comparative Analysis", paper presented at the Symposium on Public Understanding of Science and Technology, OECD, Paris.

(18) Prime Minister of Japan and His Cabinet. (2003), *Strategic Program for the Creation, Protection and Exploitation of Intellectual Property.*

(19) Prime Minister's Office. (1990), "Scientific Technology and Society", *Monthly Public Opinion Survey* August 1990 (in Japanese).

(20) Tsuchida, S. (1997), "Effects of Governmental Regulation and Information Disclosure on Public Acceptance.", *The Society for Risk Analysis: Japan section* 8 (1): 96-104 (in Japanese).

DEVELOPING COUNTRIES AND THE REGULATION OF GMOs: SOME PERSPECTIVES AND PROBLEMS.

Ruth Mackenzie[1]

Biodiversity and Marine Resources Programme

Foundation for International Environmental Law and Development

52-53 Russell Square, London, WC1B 4HP. UK.

Abstract

This contribution surveys some perspectives of developing countries in addressing policy and practical challenges associated with the regulation of genetically modified organisms (GMOs). It would be inaccurate, of course, to seek to portray developing countries as a homogenous group with regard to their priorities and concerns relating to modern biotechnology and biosafety; as in other countries, a diversity of interests and perspectives within and among developing countries gives rise to a tension in policy debates on appropriate GMO regulation. With that limitation in mind, this contribution focuses on a limited set of considerations related to the elaboration and implementation of the Cartagena Protocol on Biosafety. It surveys the main concerns and positions of developing countries in the negotiation of the Protocol and the extent to which those concerns were reflected in the Protocol as adopted. Specific areas addressed include the scope of the Protocol and its advance informed agreement procedure, the relationship with international trade agreements, socio-economic considerations, and liability and redress. The paper then considers certain implementation issues, including capacity limitations and new capacity-building initiatives; practical challenges in implementing domestic biosafety regulations; and external factors that may influence a country's policy and decision-making in respect of the design and implementation of biosafety regulation.

1. Introduction

This contribution surveys some perspectives of developing countries in addressing policy and practical challenges associated with the regulation of genetically modified organisms (GMOs). It would be inaccurate, of course, to seek to portray developing countries as a homogenous group with regard to their priorities and concerns relating to modern biotechnology and biosafety; as in other countries, a diversity of interests and perspectives within and among developing countries gives rise to a tension in policy debates on appropriate GMO regulation.

The basic tension inherent in policymaking in biotechnology and biosafety for developing countries is reflected in Chapter 16 of Agenda 21. Agenda 21 recognized the potential benefits of biotechnology in contributing to enhanced food security through sustainable agricultural processes, and called for the promotion of the development and application of biotechnologies, with special emphasis on developing countries. It further sought to promote activities to enhance biosafety regulation and international

---

[1] Parts of this paper draw upon Mackenzie, "The International Regulation of Modern Biotechnology", 13 *Yearbook of International Environmental Law* (2002). The paper also draws more generally upon the project, *Globalisation and the International Governance of Modern Biotechnology*, conducted by FIELD and the Institute of Development Studies (UK), in collaboration with the African Centre for Technology Studies (Kenya), Research and Information Systems for the Non-Aligned and Developing Countries (India) and the National Law School University of India, Bangalore. Further information on the project, which was conducted from 2000-2003, is available at www.gapresearch.org.uk.

mechanisms for co-operation, as well as enabling mechanisms for the development and environmentally sound application of biotechnology, calling for the further development of internationally agreed principles on risk assessment and management of all aspects of biotechnology. Chapter 16 of Agenda 21 noted that

> [t]he accelerated development and application of biotechnologies, particularly in developing countries, will require a major effort to build up institutional capacities at the national and regional levels. In developing countries, enabling factors such as training capacity, know-how, research and development facilities and funds, industrial building capacity, capital (including venture capital), protection of intellectual property rights, and expertise in areas including marketing research, technology assessment, socio-economic assessment and safety assessment are frequently inadequate.

On the whole, developing countries were strongly in favour of the adoption of the Cartagena Protocol on Biosafety, and they supported rather stringent safety procedures, and tended to stress issues of uncertainty, capacity, social and economic concerns, and priorities relating to food security and the protection of human health and the environment. In some contrast, in discussions in the World Trade Organisation (WTO), developing countries have tended to express general concerns about the effect of non-tariff barriers on market access for their goods. Market access concerns have thus been at the heart of the positions put forward by many developing countries in discussions in the WTO on trade and environment, on technical barriers to trade and on sanitary and phytosanitary measures, and in the context of the Agreement on Agriculture. These positions may seem difficult to reconcile, and to some degree, they may simply be reflective of the different, and often inconsistent, approaches adopted by trade, environment, or agriculture ministries in different international fora. However, they may also represent a more complex reaction to the introduction of a relatively new technology, the benefits and risks of which remain uncertain, and about which gaps in knowledge still exist in relation to the long-term impacts on the environment and on human health and in relation to the potential socio-economic impacts.

As in other countries different, often competing interests and concerns are at play in many developing countries regarding biotechnology and biosafety policy, depending upon the perspective from which the question is addressed: for example, a potential importer; a potential exporter; a potential producer; or a potential user of GMOs; or an individual, group or community which may be affected by the use of GMOs. Even setting aside broader debates around the role of modern biotechnology and other (non-GM) agricultural technologies in alleviating food security problems, biotechnology raises a series of location, environment and context-specific questions. In practical terms, some developing countries have begun to move ahead with development and use of GM technology – in particular, China and Argentina, but also, for example, South Africa, India and others.[2] Others have adopted a more cautious approach, illustrated, for example, in the response of some southern African countries to GM food aid.

The concerns expressed within and among developing countries vary: for proponents of the technology, they may relate, for example, to concerns about restricted access to GM technology and traits as a result of IPR protection, or about delays in research and development as a result of lengthy, unclear or non-existent regulatory approval processes, and the need to build public confidence through clear and effective regulation. Biosafety-related concerns focus on potential adverse impacts on the environment or on human health, and in particular on potential impacts on biodiversity, particularly in biodiversity-rich developing countries that are centres of origin of particular food crops or centres of diversity – the most prominent example of the latter to date being concerns over contamination of traditional maize varieties

---

[2] See James, *Global Status of Commercialized Transgenic Crops: 2003* (ISAAA, 2003).

in Mexico. Socio-economic concerns focus on the restrictive conditions under which GM seeds may be made available to farmers and on potential impacts of GM technology on traditional agricultural practices, replacement of traditional crop varieties, and potential loss of traditional export markets.

Given the scope and diversity of developing countries' perspectives in GMO regulation, the remainder of this contribution focuses on a limited set of issues around the elaboration and implementation of the Cartagena Protocol. Thus it focuses on biosafety, rather than on issues related to modern biotechnology more broadly, and it focuses particularly on agricultural GMOs. Section 2 highlights the main issues emphasised by most developing countries in the negotiation of the Protocol and how these concerns were (or were not) accommodated in the final text of the Protocol.[3] Section 3 considers some specific challenges related to implementation of the Protocol at the domestic level in developing countries.

2. Key Issues for Developing Countries in the Cartagena Protocol Negotiations

Developing countries were key actors in the negotiation of the Cartagena Protocol.[4] Towards the end of 1996, the African Group submitted a draft Protocol text, and, as the negotiations proceeded, a broader negotiating bloc of developing countries emerged, known as the "Like-Minded Group". This Group comprised the countries of the G-77 and China, minus the three developing countries that formed part of the "Miami Group" of major agricultural exporters, Argentina, Chile and Uruguay.[5] It has to be borne in mind, however, that the Protocol was adopted in January 2000, and negotiating positions have not necessarily remained static – with the adoption of the Protocol and its entry into force the positions of some countries have evolved, and it remains to be seen what positions they will take in future negotiations that take place regarding the future direction of the Protocol.

While developing countries played an active role in all aspects of the Protocol negotiations, central concerns focused on: the scope of the Protocol and its advance informed agreement (AIA) procedure; the establishment of strict import consent procedures for living modified organisms (LMOs);[6] the relationship of the Protocol with other international trade obligations; socio-economic considerations; liability and redress; and capacity-building.

2.1 Scope of the Protocol and of the Advance Informed Agreement (AIA) Procedure

Developing countries argued for the Protocol to be comprehensive in scope – i.e. to include all LMOs and products thereof. While the debate over "products thereof" extended throughout most of the negotiation process, products of LMOs which do not themselves consist of or contain LMOs were not finally explicitly included within the Protocol's scope. Of course, countries may still opt to subject such products to domestic regulation.

---

[3] On the Cartagena Protocol, see Mackenzie, Burhenne-Guilmin, La Vina, Werksman, Ascencio, Kinderlerer, Kummer and Tapper, *Explanatory Guide to the Cartagena Protocol on Biosafety*, IUCN Environmental Policy and Law Paper No. 46 (2003).

[4] See generally, Secretariat of the Convention on Biological Diversity, *The Cartagena Protocol: A Record of the Negotiations* (2003), prepared by the Foundation for International Environmental Law and Development, available at http://www.biodiv.org.

[5] References in the remainder of this section to "developing countries" are to the Like-Minded Group.

[6] In the Cartagena Protocol, the term living modified organisms (LMOs) is used instead of GMOs, and will accordingly be used for the remainder of this paper.

The other major controversy over the scope of the Protocol and its procedures centred on whether the Protocol should cover agricultural commodities (e.g. grains from GM crops) which, while fulfilling the definition of LMO in the Protocol, were not intended to be introduced into the environment of the Party of import. These currently constitute the vast majority of LMOs in international trade, and became know as LMO-FFPs (living modified organisms for direct use as food or feed, or for processing).

The Miami Group and the biotechnology industry argued that, since they were not intended for introduction into the environment, LMO-FFPs should not be covered by the Protocol, which was intended primarily to address potential risks to biological diversity. Developing countries argued for their inclusion, on the basis that, notwithstanding their intended use at the point of export, in developing countries grain imports intended for food, feed or processing could end up being planted in the environment deliberately, or could be accidentally released into the environment (e.g. through spillage), and hence they should be treated in the same manner as other LMOs.

The concerns of developing countries were only partially met here, with LMO-FFPs included within the Protocol's scope, but excluded from its AIA procedure. Thus, there is no direct obligation in the Protocol to notify a Party of import of an intended transboundary movement of an LMO-FFP. Instead, the Protocol incorporates, in its Article 11, a new multilateral information exchange procedure, coupled with the right to impose domestic regulation of LMO-FFP imports. Thus, developing countries may select to subject imports of LMO-FFPs to risk assessment and approval procedures prior to import, similar to the AIA procedure, but it is their responsibility to make exporters aware of their relevant regulations through the Biosafety Clearing House, and to check the Biosafety Clearing House to keep informed of what LMO-FFPs may be subject to transboundary movement. This poses the problem for developing countries of tracking information on new products through the BCH (see below) and putting an appropriate regulatory system in place.

2.2 Strict AIA Procedure

The core procedural mechanism in the Protocol is the advance informed agreement procedure. Throughout the negotiations, developing countries' positions sought to ensure that the coverage of the AIA procedure was broad, and that it incorporated a requirement of explicit consent to imports, and adequate time frames for decision-making by developing countries. In the event, the scope of the AIA procedure is rather narrow: it covers only the first intentional transboundary movement of an LMO for intentional introduction into the environment of the Party of import. In addition to LMO-FFPs, LMOs destined for contained use and LMOs in transit are excluded from the AIA procedure under the Protocol, although countries remain free to regulate them at the domestic level (Article 6).

However, the AIA procedure does reflect the concerns of developing countries in respect of explicit consent requirements and timeframes. It provides for a period of up to 90 days within which a Party of import must acknowledge receipt of a notification of a proposed transboundary movement of an LMO, and a 270 days period for reaching a decision on import, with the possibility of extension for a defined period of time, and the possibility to "stop the clock" where further information is requested (Articles 9 and 10). In addition, it states that failure to respond to a notification or to communicate a decision within the stated time periods shall not imply a consent to the transboundary movement. In the light of the capacity constraints discussed in section 3 below, it remains to be seen whether the time periods in the Protocol will provide adequate time in practice to reach an informed decision.

Other aspects of the Protocol's AIA procedure designed to meet concerns of developing countries expressed in the negotiations are the possibility for the Party of import to require the exporter to meet the costs of risk assessment. In addition, developing countries pushed strongly for the inclusion of the precautionary principle in the operational provisions of the Protocol on decision making on imports – allowing Parties of import to prohibit or restrict imports where there is a lack of scientific certainty over the extent of potential adverse effects of an LMO. This concern is now addressed in Article 10(6) of the Protocol (and in Article 11(8) in respect of LMO-FFPs).

2.3 Relationship between the Protocol and International Trade Rules

The relationship between the Protocol and other international trade rules established in the WTO agreements has been considered in detail in another contribution at this workshop.[7] The Like-Minded Group strongly opposed the inclusion of any savings clause in the Protocol that would have accorded primacy to international trade rules over the rules and procedures established by the Protocol. The three preambular paragraphs of the Protocol which address now this relationship do not firmly resolve the issue. Indeed, as discussed further below, the relationship between the Protocol, domestic biosafety measures and WTO rules remains somewhat uncertain, giving rise to difficult policy choices for developing countries as they implement the Protocol.

2.4 Socio-economic Considerations

Much of the debate around the role of biotechnology in developing countries has focused on potential socio-economic impacts of LMOs, in particular, on the conditions under which GM seeds might be made available (for example, licensing arrangements and impacts on the ability of farmers to save and re-use seed) and on the potential impacts of GM crops on traditional crop varieties. Concerns have also been expressed that developments in modern biotechnology could give rise to substitutions of crops traditionally exported from developing countries, thus reducing export revenues. In some circumstances, the potential impacts of a particular LMO upon biological diversity, plant or animal life or health, or human health may give rise directly to related food security or socio-economic concerns. In others, socio-economic concerns not specifically related to direct impacts on biodiversity or health might be raised by the potential use of GM crops.

Initial proposals from developing countries in the Cartagena Protocol negotiations, particularly from the African region, sought wide-ranging provisions on the role of socio-economic considerations in decision-making on imports of LMOs. For example, an early proposal listed the parameters of a risk assessment to include the following socio-economic considerations:

(1) anticipated changes in the existing social and economic patterns; possible threats to biological diversity, traditional crops or other products, and, in particular, farmers' varieties and sustainable agriculture;
(2) impacts likely to be posed by the possibility of substituting traditional crops, products, and indigenous technologies through modern biotechnology outside of their agro-climatic zones;
(3) anticipated social and economic costs due to the loss of genetic diversity, employment, market opportunities, and, in general, means of livelihood of the communities;
(4) disruptions to social and economic welfare; and
(5) possible effects contrary to the social, cultural, ethical and religious values of communities.[8]

---

[7] See paper by Zarilli.
[8] See UNEP/CBD/BSWG/2/2.

The breadth of this proposal was unacceptable to most developed countries, although some, such as Norway, were prepared to see socio-economic concerns reflected in the Biosafety Protocol in some way. Other countries, such as Bolivia and Mexico, emphasised the importance of economic impacts on centres of origin and genetic diversity.

The debate led to the inclusion of Article 26 in the Protocol, paragraph 1 of which provides that:

> The Parties, in reaching a decision on import under this Protocol or under its domestic measures implementing the Protocol, may take into account, consistent with their international obligations, socio-economic considerations arising from the impact of living modified organisms on the conservation and sustainable use of biological diversity, especially with regard to the value of biological diversity to indigenous and local communities.

The scope of Article 26 remains to be clarified, both in national legislation and in practice, and perhaps in further elaboration by the meeting of the Parties to the Protocol. While Article 26 would appear to extend to the impacts on local crop varieties used by indigenous and local communities (and potentially to related, but indirect, impacts), it is more debatable whether it would extend to more general socio-economic impacts of the use of GMOs in a country—for example, such as what happens to local/national food production patterns as a whole and how this affects different groups. In addition, the phrase "consistent with their international obligations" indicates that a country's conduct in relying upon Article 26 can be tested against other applicable international agreements, including WTO agreements.

2.5 Liability and Compensation for Damage

Another issue of key concern to developing countries throughout the negotiations concerned who should bear the risk of any damage caused by LMOs, and the extent to which specific rules and procedures on liability and redress should be incorporated into the Protocol. This question gave rise to considerable debate and disagreement during the negotiations. Some developing countries proposed strict (no fault) liability of Party of export for any damage caused by an LMO, or, alternatively, strict liability of the exporter. They argued that existing rules of international law were inadequate to ensure redress for any damage that may be caused by LMOs and sought to include specific rules within the Protocol. Other countries, and the biotechnology industry, expressed the view that existing general domestic rules on liability for damage to persons, property or the environment, including product liability rules, were an adequate and appropriate means to address any question of damage arising from LMOs, and that LMOs did not pose special risks meriting specific national or international liability rules or procedures. Others, on the basis of experience of other international negotiations on liability issues, took the view that this was a matter that simply could not be resolved within the timeframe set for the completion of the Protocol negotiations.

In the event, no final decision was taken on this issue and the Protocol incorporates an enabling provision requiring the Parties to the Protocol to consider this issue at their first meeting and to establish a process with a view to considering appropriate international rules and procedures with a view to resolving the question within 4 years of the entry into force of the Protocol. It seems likely that a working group will be established by the first meeting of the Parties in February 2004 to undertake further work on this contentious issue. In the meantime it remains to be seen whether countries will seek to integrate specific liability rules into their domestic biosafety frameworks.

3. Implementation Issues

A number of factors will affect the ability of developing countries to implement the Cartagena Protocol effectively in a manner that meets their national concerns and priorities related to the safe use of modern biotechnology. These include issues of capacity, which remain high on the international agenda, but also other practical problems concerning the implementation and enforcement of biosafety regulations, and the broader political context within which national biosafety regulatory frameworks are being developed and implemented.

3.1 Capacity-building

The Biosafety Protocol is premised upon a system of national biosafety frameworks and upon information exchange. National biosafety frameworks need to be capable, *inter alia,* of dealing with applications for the import and use of LMOs, assessing risks associated with LMOs, policing "illegal" transboundary movements and unauthorized use of LMOs, managing the risks identified, and monitoring the actual impact of authorized LMOs in the receiving environment. It is far from clear that the situation on the ground in many countries will enable these requirements to be met.

During the negotiations of the Protocol, developing countries pressed for provisions on capacity-building in biosafety and biotechnology, as well as the provision of financial resources for the implementation of the Protocol. Article 22 provides for Parties to cooperate in the development and/or strengthening of human resources and institutional capacities in biosafety, including biotechnology to the extent that it is required for biosafety. Key areas of capacity-building would appear to be legal and institutional; risk assessment and risk management; information management; and public awareness and participation.

In terms of legal and institutional capacity, the Intergovernmental Committee for the Cartagena Protocol (ICCP) has developed a checklist of legal and administrative requirements for implementation of the Protocol.[9] In general terms, Parties need to establish a competent national authority (or authorities) to receive notifications of proposed imports of LMOs; they need to be able to respond to those notifications and to reach decisions based on risk assessment and any other relevant considerations. In order to do this, a domestic regulatory framework will be required setting out which institution(s) will be responsible for dealing with such notifications and making decisions, and how and on what basis decisions will be made. Such frameworks are not yet in place in many developing countries. While many countries have some relevant laws, regulations or guidelines in place, often these have been developed in a piecemeal fashion, and only address a limited set of activities involving LMOs, such as contained use or field trials, or address some aspects of biosafety in the broader context, for example, of plant protection/quarantine regulations or environmental impact assessment requirements. Few presently address biosafety in a comprehensive manner, covering issues such as authorisation of imports, exports or commercial use of LMOs, or identification or labelling requirements.

In addition to establishing the basic legal framework for regulation of LMOs, it is evident that significant capacity-building is required in relation to scientific and technical aspects of biosafety – particularly in the fields of risk assessment and risk management. In existing regulatory frameworks for biosafety, input from a wide range of scientific disciplines may be called upon in the risk assessment process. By way of example, the UNEP International Technical Guidelines for Safety in Biotechnology indicated a list of various fields of scientific expertise that might be required for the conduct of risk assessment, including molecular genetics, ecology, microbiology, taxonomy, agronomy, epidemiology and so on. The

---

[9] ICCP Recommendation 3/5, Annex III.

capacities of developing countries in these and other relevant fields clearly vary greatly, but for some, at least, significant efforts will be required to ensure that sufficient inputs are available for risk assessment to underpin decision-making under the Protocol, and to implement and monitor risk management measures. Such limitations may lead to other, seemingly unrelated, problems – for example, where capacity is limited, conflicts of interest may arise where those who are conducting biotech research are called upon, by virtue of their relevant expertise, also to act in some regulatory capacity.

Some mechanisms are being put in place under the Protocol to provide assistance to countries, upon request, to facilitate decision-making - for example a roster of experts has been established to "provide advice and other support, as appropriate and upon request, to developing country Parties and Parties with economies in transition, to conduct risk assessment, make informed decisions, develop national human resources and promote institutional strengthening, associated with the transboundary movements of living modified organisms".[10] However, such measures are by no means a long-term substitute for the development of capacity at the national level. Other measures that countries may choose to consider include a degree of regional or sub-regional cooperation in relation to risk assessment or risk management.

As noted in Section 2 above, implementation of the Protocol will rely heavily on the effective establishment and utilisation of the Biosafety Clearing House (BCH). As well as its pivotal role in relation to information-sharing on LMO-FFPs, the BCH will include information on national laws and regulations; bilateral and regional agreements covering transboundary movement of LMOs; summaries of risks assessments and environmental reviews generated by domestic regulatory processes in respect of specific LMOs; and final decisions of countries of imports with regard to specific LMOs. While the Protocol makes some provision for countries which have limited internet access, it is clear that the BCH will operate primarily as an electronic/internet-based resource that will require significant input from Parties. Effective implementation of the Protocol will therefore require significant resource- and capacity-building efforts towards enhancing reliable access to the BCH, and in building adequate human and technical capacity for information management in order to ensure that the BCH is kept up-to-date.

3.1.1 Capacity-building Initiatives

As might be expected, capacity-building has featured high on the agenda of the Intergovernmental Committee for the Cartagena Protocol in the period since the Protocol's adoption. The ICCP has endorsed a capacity-building action plan, and overseen the establishment of the roster of experts. In addition, in the wake of the adoption of the Protocol, a raft of new bilateral and multilateral capacity-building initiatives has been launched aimed at facilitating implementation of the Protocol. The most wide-ranging in terms of country coverage is a Global Environment Facility (GEF) funded project, being implemented by the UN Environment Programme (UNEP), to assist countries to develop national biosafety frameworks (NBFs) to implement the Protocol. More than 100 countries (including countries with economies in transition) are now participating in this project. Its aims are limited to the development of national biosafety frameworks. Implementation of that framework is seen as a subsequent, and separate, step and eight countries which participated in a pilot project to develop NBFs are now participating in separate implementation projects. While the project focuses on the development of a national regulatory framework for biosafety, it also pays particular attention to risk assessment and risk management issues and to public awareness and participation.[11] The project is running a series of

---

[10] The roster was established by the Extraordinary Meeting of the Conference of the Parties to the Convention on Biological Diversity upon adoption of the Protocol, in decision EM-I/3.

[11] Information on the project can be found at http://www.unep.ch/biosafety.

regional and sub-regional workshops to support national level activities. The success of this initiative in the long run, however, will be judged not on whether countries put regulatory frameworks in place on paper, but on whether it is followed up by concrete activities aimed at implementation in practice. Further financial support for implementation may be available through GEF medium-size projects, and through the provisions of the Protocol on financial resources.

In addition to this effort, numerous bilateral initiatives are underway with similar, although not identical, objectives. Coordination of these initiatives, together with related initiatives of other multilateral agencies such as FAO, is likely to be another significant factor in effectiveness. It is evident that any credible efforts to build capacity in risk assessment and risk management need to be long-term, and one question which arises is what interim measures developing countries might apply pending fruition of the current raft of capacity-building initiatives. This issue is not answered by the Protocol, but might, it could be argued, justify a precautionary approach.

3.2 Practical problems: biosafety and the "governability" of the seed trade

While the Biosafety Protocol is designed to ensure the safe use of LMOs and sets out reasonably clear rules to cover imports of LMOs intended for intentional introduction into the environment of Parties of import, it is sometimes easy to lose sight of what is actually being regulated in practice. It is important therefore to consider what practical limitations exist on the "governability" of the seed trade.[12] Experience to date, in developed as well as developing countries, suggests that even with applicable legal frameworks in place, effectively regulating agricultural LMOs is a challenging task. There have now been numerous incidents reported of accidental release and illegal growing and trading in seeds which governments appear to have been unaware of. Examples include the illegal sale and growing of *Bt* cotton in Gujarat state, India; the informal exchange of *Bt* cotton and non-GM seeds in China; the apparent contamination of traditional maize varieties in Mexico originating from GM maize; unauthorised transboundary movements of LMOs into Brazil and Zimbabwe; concerns expressed by some countries in southern Africa about controlling the use of seeds entering a country as food aid; the StarLink experience in the United States, and instances of unauthorised releases of LMOs in the UK. While difficulties associated with the effective implementation, monitoring and enforcement of environmental laws are familiar in most jurisdictions, each incident of this nature highlights the additional problems associated with regulating intra- and inter-state trade and movements of a commodity as difficult to regulate as seeds.[13]

As noted in Section 2, one of the principal concerns of developing countries regarding the exclusion of LMO-FFPs from the Protocol's advance informed agreement procedure was that such LMOs may well find their way into the environment of the Party of import, whether accidentally, through spillage, or through deliberate planting. These concerns are exacerbated by weak identification requirements for LMO-FFPs in the Protocol, and practice of commingling GM and non-GM grains in LMO exporting countries. Monitoring and detection of LMOs therefore raises significant challenges for developed and developing countries – what techniques are available for testing whether or not a shipment contains LMOs, what LMOs it contains and in what proportion?

---

[12] See Mackenzie and Newell, *Globalisation and the International Governance of Modern Biotechnology: Promoting Food Security?*, available at http://www.gapresearch.org/governance/
[13] Mackenzie and Newell, note 12 above.

3.3 Trade and Trade Politics

While the Protocol sets out procedures for imports of certain LMOs, it does not prescribe the circumstances in which a country should accept or reject the import of a specific LMO. There is therefore plenty of scope for disputes in the application of the Protocol procedures to specific LMO imports. Countries developing and implementing national biosafety frameworks are likely to be subject to an array of influences and pressures, both internal, from interest groups and stakeholders, and external, from countries that wish to export LMOs to developing countries, or those that constitute markets for developing country agricultural exports, or from bilateral or multilateral donors. The leverage of external actors is increased by the dependence of many developing countries on trade and aid. Several examples have been publicised over recent years of developing countries being subject to strong forms of political pressure to revise their biosafety regulations in line with the interests and concerns of more powerful exporting countries. Bolivia, Sri Lanka, Egypt, China and Thailand have all reportedly been subject to such pressures to relax the stringency of their regulatory systems or to retreat from *de facto* moratoriums on the technology's import and development within the country. Such pressures are often, it seems, exerted at high level.[14] Indirect pressure may also be exerted as a result of concerns about marketability of products – for example, if consumers in a country's primary trading partners reject GM products.

The uncertain relationship between the Biosafety Protocol and the WTO also exposes developing countries to difficult policy choices. There remains a degree of uncertainty and unpredictability regarding the scope for countries take into account socio-economic considerations in decision-making on imports of LMOs. Uncertainty and ambiguity still exist with regard to how domestic environmental and health measures should be designed and applied if they are to be deemed to be consistent with the relevant WTO agreements. This situation arises out of ambiguities within the WTO agreements themselves, uncertainties regarding the relationship between the Biosafety Protocol and the WTO and between the WTO and MEAs, the lack of international consensus on the benefits and risks of LMOs in different contexts, and on how to accommodate different views and circumstances in international regulation. While the uncertainty and ambiguity in the international legal framework on LMOs may present some flexibility, autonomy and space for countries as they develop and implement their own national biosafety frameworks, it may also present certain risks. Unpredictability in the outcome of possible disputes surrounding domestic LMO regulations may render developing countries more vulnerable to additional pressure from powerful trading partners and give rise to delayed or weakened national regulations.

4. Conclusions

Developing countries are adopting different approaches to the regulation of LMOs and are at different stages of researching and using modern biotechnology, and of establishing and implementing regulatory frameworks for biosafety. At the most general level, however, developing countries face the same challenges as any other in addressing appropriate regulation of LMOs, such as:

- ensuring societal trust and confidence in regulation;
- integrating concerns of different stakeholders into the regulatory and administrative framework for biosafety;
- undertaking risk assessment;
- dealing with scientific uncertainty; and
- complying with various international obligations.

---

[14] Mackenzie and Newell, not 12 above.

In addition, developing countries face often severe capacity constraints, in terms of establishing or adapting institutional infrastructure for regulation of LMOs, in building public awareness, and in assessing and managing risks associated with LMOs. These capacity constraints have been emphasised in international fora addressing LMO regulation, and are now the subject of numerous capacity building efforts.

The elaboration or revision of national biosafety frameworks, which many developing countries have recently initiated, represents an important opportunity to consider and address many of these issues. It is critical that these processes and their outputs are comprehensive, deliberative, and responsive to the complexity of national circumstances and interests. External pressures should not foreclose real debate about policy directions with public consultation for the establishment of workable and effective biosafety systems.

SESSION 4:
LESSONS LEARNED AND
REMAINING CHALLENGES

SCIENTIFIC CHALLENGES FOR RISK ASSESSMENT

Hubert P.J.M Noteborn and Wim de Wit
Dutch Food and Consumer Product Safety Authority (VWA).
Directorate Research and Risk Assessment,
P.O. Box 19506, 2500 CM The Hague, The Netherlands.

Abstract

Related to food safety aspects, the concept of substantial equivalence is a guiding tool for the assessment, developed by OECD and further elaborated by FAO/WHO. One of the safety concerns regarding GM crops pertains to the potential of unintended effects caused by processes of transgene insertion (DNA re-arrangements) or from metabolic effects of novel gene product(s). Unintended effects are phenotypic, response or compositional effects which go beyond that of the original genetic modification and which might impact primarily on health. Unintended effects occur in both GM and non-GM crops; however, GM crops are better characterised. Conventional 'targeted' analytical approaches and animal feeding trials with whole foods may reveal unintended effects only by chance or if anticipated. Therefore, it is a scientific challenge to develop new methodology that allow for a 'holistic' simultaneous screening of potential alterations in the physiology of the GM crop at different biological integration levels. The present tendency is to include metabolomics (parallel analysis of a range of primary and secondary metabolites), proteomics (analysis of polypeptide complement) and transcriptomics (parallel analysis of gene expression). Current experiences suggest that the safety assessment of GM crops should focus primarily on the intended novel traits. Whereas the 'omics' technologies should not at present be an official requirement, but their further development should be actively encouraged. Concerning the search for unintended effects as part of the overall risk assessment framework its relevance might be questioned. In other words, do we really reduce the uncertainties in the hazard identification and risk assessment by collecting huge numbers of data?

1. Introduction

Some foods have been consumed over centuries, while others have been introduced to Europe, for example, from the New World (maize, potato, tomato) or Asia (rice, spices, soybean). Therefore, we do know a lot about our food, and the safety of conventionally bred crops is taken for granted based on our history of safe use. Which means that mankind has trusted for centuries in the fields and in the kitchen on the principle of '*pass it on from father to son and from mother to daughter*'. But at present and beyond that, we must trust risk assessors who demonstrate that GMO-based food is as safe as its traditional counterpart. In the USA, the commercial cultivation of genetically modified (GM) crops with benefits to farmers (e.g. pest resistance, herbicide tolerance) has increased substantially since their market introduction in the mid-1990's [1]. However, the commercialisation of GM crops in Europe is practically non-existent at the present time. Moreover, future GM crops are envisaged that will also provide health benefits to their end-users: the consumer. In Europe, the introduction of crops produced by genetic engineering methods onto the market place is regulated under the Novel Foods Regulation [2]. To assess the food safety of GM crops internationally recognised strategies have been developed. As a result, a thorough pre-market risk assessment is required. Related to GM crops and derived products the risk assessment process pays particular attention to potential adverse effects that might compromise human and animal health and environmental biosafety. The food safety hazards pertaining to GM crops are:

i)   toxicological properties of new gene products and transfer via animal feed to edible animal-derived products;

ii)   formation and toxicity of new metabolites of herbicides and or insecticides applied to GM crops and possible changes in their residue profiles and contents;

iii)   gene transfer of plant DNA to the gut microflora;

iv)   potential allergenicity, and;

v)   unintended effects, which represents one of the main issues in the safety assessment of GM crops, and a scientific challenge too.

It is noted that this vigorous testing is not a requirement for the introduction of novel seed varieties bred by conventional breeding practices. As a result GM crops are therefore better characterised than ever has been done before in case of conventionally bred crops, including knowledge on the site and nature of the genetic modification. Furthermore, the European Commission intends to tightening up the regulatory process by addressing the concerns of its citizens and Member states and to pave the way for removing the current *de facto* moratorium (i.e. concept-Regulation on Traceability and Labelling).

This paper deals with the risk assessment of unintended effects in GM crops i.e. effects which go beyond that of the original modification and which might impact primarily on health of man and animal. Effects that could evoke hazardous perturbations in gene expression, metabolic pathways and or metabolism that affect its food, feed or nutritional status. Especially, the scientific challenges and measures to detect unintended effects in GM crops will be highlighted and critically discussed.

2. Substantial equivalence

The aim of the risk assessment of foods and feeds produced by GM technologies is to demonstrate that the novel crop, food or feed is as safe as its traditional counterpart (where one exists) and as such does not introduce any additional or new risks to the health of man and animal. This approach involves the concept of *Substantial Equivalence*. This concept is a comparative analysis and embraces the idea that the existing traditionally produced food supply can be considered as safe due to its long history of safe use. Substantial equivalence has been formulated as a guiding tool for the assessment of GM foods as part of a general safety evaluation framework [3-6]. Which means that the characteristics of the modified crop are compared to an existing traditionally bred crop. With respect to plants, this is most usually the parent crop from which the genetic modifications were made. It is considered a useful concept to perform a relative risk assessment of GM crops and has proven to be adequate for the assessment of GM crops now admitted to the market place. The properties of the GM crop have been compared with their appropriate counterparts, even though it has been recognised that current foods may also contain many anti-nutrients and toxicants, which at certain levels of consumption, may induce adverse effects in humans and animals. The challenge here is to correlate a phenotypic modification (e.g. stunted growth) with alteration in chemical composition.

The concept of *Substantial Equivalence* is fully endorsed in the European regulatory framework [2]. It is stated that the assessment of substantial equivalence is an analytical process, where the novel food is compared to the most appropriate approved food, not necessarily meaning a conventional food, but possibly an earlier approved genetically modified variety too. It should be borne in mind that it is not a safety assessment in itself i.e. there is not a toxicological endpoint *per se*. The comparative analysis does not characterise the hazard and or risk. Which means that substantial equivalence is a starting point in the safety evaluation of GMOs, rather than an endpoint of the risk assessment. As stated before, its application assists to identify the similarities and differences between an existing, conventionally produced, food and the new GM-product, which are then subject for further toxicological investigations, if required.

Notwithstanding the fact that the concept of *Substantial Equivalence* has been adopted at the regulatory level, the EU consumer and or citizen are not fully convinced that risk assessors and risk managers of governments, industry, food authorities and inspection services do a proper job in the field of GMOs [7,8].

3. Practical Implications of Substantial Equivalence

In general, three scenarios are envisioned in whom the GM-crop or -food would be:
i)   substantially equivalent;
ii)   substantially equivalent except for the inserted trait, or;
iii)   not equivalent at all.

In the first scenario, no further specific toxicological testing is required since the product has been characterised as substantially equivalent to a traditional counterpart whose consumption is considered to be safe, like starch from GM-potatoes [9]. In the second scenario substantial equivalence would apply except for the inserted trait, and therefore, the focus of the safety testing is on this trait, like in case of a genetically modified BT-tomato [10]. Following an extensive characterisation of the recombinant BT-protein, its safety can be demonstrated using a case-by-case strategy within a tiered approach according to the nature and function of the newly expressed protein (*i.e.* biopesticide). The issue of the potential occurrence of adverse compositional effects due to the genetic modification process, such as, the loss of existing traits or the acquisition of new ones, was examined too. Thereto, a subchronic animal feeding study of 90 days' duration according to OECD guideline 408 was considered to be the minimum requirement to demonstrate the safety of repeated dietary intake of a GM food. In the third scenario, the GM-crops will be not equivalent at all and, it is speculated that the chance of unintended effects might increase too. For example, through the method of metabolic pathway engineering GM crops will contain compounds with health benefits for their consumers [11]. These newer genetic alterations, changing agronomic or nutrition-related properties, are more complex, involving profound changes in metabolic pathways as a result of the intended genetic modification. Plant compounds, like carotenoids, (iso-)flavonoids and phytosterols, have been identified to lower, for instance, the incidence of certain cancers and cardiovascular diseases. Although, the making out of health claims is still a major scientific challenge, these intrinsic plant constituents are considered as very attractive (new) targets for genetic engineering, and as a consequence thereof, levels of b-carotene or flavonoids have been up-regulated in rice [12] or in tomatoes [13,14]. On the other hand, screening for potential undesirable changes in these GMOs becomes a scientific challenge as, for instance, an appropriate parent (counterpart) may not exist anymore (i.e. worst-case scenario).

4. Sense of Reality of Unintended Effects

Characterisation of GM crops is a legal requirement, and the hazard of unintended effects is part of the safety assessment. As a basis for this concept definitions have been adopted [4,15,16]. Intended effects of genetic engineering are those which are targeted to occur from the introduction of the gene(s) in question and which fulfil the original objectives of the genetic transformation process. Whereas an unintended effect is a statistically significant difference in the phenotype, response, or composition of the GM crop compared with its parent from which it is derived and, which does not fulfil the original objectives of the genetic modification by taking the intended effect of the inserted or silenced target gene into account. With respect to unintended effects two additional definitions have been formulated and adopted:
i)   predictable unintended effects, and;
ii)   unpredictable unintended effects.

Predictable unintended effects are those unintended changes in the GM crop that go beyond the primary expected effect(s) of introducing target gene(s), but which may be explicable in terms of current knowledge of plant biology and metabolic pathway integration and interconnections. Contrarily, unpredictable unintended effects are those unexpected changes which fall outside the present level of understanding of plant biology and physiology. As a result, predictable and unpredictable unintended effects may, or may not prove to have relevance in terms of the GM-crop and derived product safety, but must be taken into account when assessing the risk. Which means that unintended effects may have positive, negative or indeed no consequences on the agronomic vigour or safety profile of the GM-crop. Thus, unintended effects do not automatically infer health hazards. Ideally, only those parameters which fall outside the range of natural variation should be considered further in the hazard identification and risk assessment.

5. Frequency of Unintended Effects

As a part of agriculture, man started rearing plants to meet his requirements. This is when humans started to learn how to influence the process of natural evolution so as to breed crop plants. Since that time, plant breeding exploits the natural molecular pathways of DNA exchange and repair by using the natural genetic variation in combination with artificial selection or inducing new variability via artificial means, like the use of mutagenic chemicals and irradiation. It appears that the non-homologous end joining (NHEJ) process is the predominant (repair) form of recombination in plants [17]. Whereas Britt [18] indicated that the subsequent double strand break (DSB) repair in plants is more error prone than in other living organisms. Therefore, errors that change the original sequence occur at very high frequency in non-GM plants. It is also known that gene-rich regions, that are transcriptionally active, are the preferred sites of natural DNA recombination. However, why genes *per se* are 'hot spots' for recombination is not well known as yet [19]. Due to innovations of the recDNA-technology exogenous DNA can be integrated into the plant'genome [20]. Two commonly used methods of transgene DNA delivery are the biolistic or microprojectile bombardment system and the *Agrobacterium tumefaciens*-mediated transformation. Related to these 'gene-transfer' systems, it is inherently impossible to predict the fate and site of a particular integrated transgene construct into the plant genome, given its nucleotide sequence. Thus, a possible consequence of random integration of transgenes in the crop's DNA can be the disruption of endogenous gene functions due to the insertional mutagenesis process. Given the aforementioned hazard, two questions arise for the risk assessor:

i)   Is transgene integration in plant chromosomes any more likely to result in DNA disruption compared to the natural recombination mechanisms?
        and
ii)   Is there an increased chance of a transgene being integrated into active gene-rich regions compared to the other chromosomal locations in the crop?

Reported data indicate that the random integration of transgenes into the plant genomic DNA is identical to the preferred recombination mechanism that occurs in all plant cells, especially during mitosis. Transgene integration uses the natural ability of plant cells to exchange genetic material during DNA duplication by NHEJ. As natural NHEJ is an error-prone pathway that introduces deletions and new filler DNA into the recombination zone it results in DNA patchwork, which has also been observed for transgene integration sites. There is an increased chance of 'hot spot' integration, because exogenous genes preferentially integrate into transcriptionally active regions of chromosomes [21]. But transgene DNA apparently behaves not differently from that of endogenous plant DNA. Therefore, the introduction of selected characteristics (i.e. the intended effects) might also be a source of potential unintended effects both in conventional and gen-biotech plant breeding. However, the order of magnitude of the frequency (risk) is unknown, and apparently case dependent. To estimate this frequency is therefore an other scientific challenge.

6. Evidence of Unintended Effects

Since JG Kolreuter' publication in 1761, hybridisation followed by selection has been the major tool of conventional plant breeding. After a variable population is recognised, individuals that are the best performers for the desired trait (e.g. yield, size or quality) are chosen and the rest of the population is discarded or rejected. The progeny of selected individuals is grown further and again screened for the desired feature. This process (i.e. multi-generation selection) is repeated until a uniform plant population is attained, which has the best-desired characters. Eventually, a desired uniform crop variety is produced by this successive selection and backcrossing. The hallmark of the selection lies in the breeder's ability to choose the best and safest plants from a cluster of many, however, criteria are highly dependent on his skills and expertise. The extensive backcrossing procedures, that takes many years, remove undesirable phenotypes, and can therefore be considered as a safety guard. It should kept in mind that the same field selection processes apply to both conventional and GM breeding. Perhaps, due to the hard selection of favourable lines and discarding those having undesirable features, there are only a few, extremely rare, cases reported in literature where unintended effects have given rise to safety concerns. These cases were identified once the non-GM crop was already on the market. Examples can be found for pest resistance breeding: such as, a celery with a high content of furanocoumarins [22], potatoes with increased levels of glycoalkaloids [23] and squash/zucchini showing harmful levels of cucurbitacin [24]. There are no indications that unintended effects are more likely to occur in GM crops, however. Examples of unexpected secondary effects due to somaclonal variations, pleiotropic effects or even genetic modification that may be of biological or agronomic importance to the plant have been summarised in a previously published paper (for details see [25]). Some of these alterations would be indicative that the experimental GM plant does not possess the appropriate properties to allow a further development into a commercial crop [26]. Others would only be identified through appropriate trials in the field, like the well-known soybean-case published in 1999 by Gertz and co-workers [27].

Related to the aforementioned examples, the challenge is to discriminate between environmental factors and those caused by genetic modification. To assure that unintended effects of the genetic modification are identified by more scientific measures, from the mid 90s until now, two different strategies have been applied: i) a targeted compositional evaluation i.e. a profile of major nutrients and known toxicants and ii) the testing of the whole GM crop or food in laboratory animals.

7. Targeted Analysis

The result of the targeted analytical approach heavily depends on the analytes selected for the comparison. An extensive chemical analysis of key nutrients, antinutrients and toxicants typical for the genetically modified Bt-tomato plant compared to its counterpart and a 'normal range' of that crop has been reported earlier [10]. This approach has been further evaluated in the EU 5FP RTD project GMOCARE [28], which workplan concentrated both on a range of transgenic potato lines modified in their starch composition, (defective) glycoprotein processing, polyamine-, sugar-, glycoalkaloid- or amino acid-metabolism and transgenic tomato lines with elevated levels of phytosterols and or isoprenoids (carotenoids). The GM-crops were continuously characterised on sizes of T-DNA, vector backbone integration, copy number and expression levels. Morphological phenotype (i.e. growth, plant architecture, number, shape and colour of leaves, flowers and tubers, precocity) was checked and compared to observations from previous cultures. The targeted analyses of proximates (e.g. protein, fat, moisture, ash) as well as selected key compounds indicated that despite some very significant phenotypic differences associated with specific constructs (e.g. stunted growth, modified numbers of tubers) no transgenic potato line was consistently different from the wild type control in any selected metabolites measured to date. There were no observations of deleterious effects on the physiology of the GM tomato plants, which were modified in their carotenoid biosynthesis of which the carotenoid (phytosterol) profiles were significantly changed [13]. With transgenic potato lines developed to modify the glycoalkaloid metabolism, a significant reduction in solanine content has been obtained. On the contrary, total glycoalkaloid levels appeared to be unaffected due to a corresponding increase in the other major tuber glycoalkaloid, chaconine. This provides excellent evidence for metabolic compensation in transgenic lines (Shepherd et al. submitted for publication). It was also observed that the degree of phenotype exhibited can change from one progeny to another. Therefore, the following question has been raised: does the substantial equivalent change during inheritance?

With respect to the selection of analytes to be examined consensus documents on compositional considerations for new varieties (i.e. key food and feed nutrients and antinutrients) have been written for soybean and low erucic acid rapeseed (canola). While monographs on corn, potato, sugar beet and rice are in progress [4]. Limitations to this targeted (single compound) analytical, comparative, approach are, however, the possible occurrence of unknown toxicants and antinutrients, in particular, crop species with no history of (safe) use. Beside it, the availability of adequate detection methods for all compounds does not exist. Generally, there are also limited data on the agronomic and compositional properties and, on the natural background in constituent levels (i.e. natural variation or base-line properties). There is a lack of information on the natural variation within and between given plant cultivars for all the parameters that can now or in the future be measured. Which means that acceptable degrees of compositional differences between a GM-crop and the range of that crop can not be defined, given also the great dispersal in published food composition of macronutrients and micronutrients, say for tomato in case of vitamins like b-carotene ([29]; http://www.fao.org/infoods/tables_europe_en.stm). There is an urgent need for a clear definition of a plant's composition including background variability. In addition, criticisms of this 'single-analyte' strategy are that it is open to bias and will never pick up an unintended effect. With regard to this there are gaps related to:

i)   availability of sufficient knowledge of plant biology and metabolic pathway integration and interconnectivities;
ii)   availability of a comparator with a similar genetic background;
iii)   availability of a few 'miracle' compounds to describe the whole complexity of crops.

8. Animal Feeding Trials

The minimum requirement to demonstrate the safety of long-term consumption of a food is a sub-chronic 90-days study (OECD guideline 408). To ensure that the semisynthetic diet containing the whole food is palatable to the test animals a short duration, 28 days, study according to OECD guideline 407 is also a prerequisite [10]. Testing of whole foods in laboratory animals has its specific problems too, and considerable experience has been gained with the toxicological testing of irradiated foods. Feeding animals with whole foods at exaggerated dose levels may induce a series of adverse effects, which would mask potential adverse effects caused by unintended effects induced by the genetic modification [7]. Thus, there are very limited possibilities to test a wider dosage range than possible by the straightforward feeding tests with additives, pesticides and residues [30]. Due to their bulk and specific composition, confounding results may result of the intrinsic toxicity of plant constituents, when tested as single compounds, or result due to matrix effects, since crops contain a myriad of components. In case of whole

GM-crops it is postulated that a No Observed Adverse Effect Level (NOAEL) cannot be estimated due to the compositional complexity and, most important, based on the fact that certain plant constituents show both beneficial and adverse effects depending on their dose-range applied.

9. Non-targeted Analysis

As a result, it is therefore desirable to develop new methodology, which will allow for the simultaneous screening of potential changes in the physiology of the GM-crop as was recommended by OECD via a response to the G8 2000 meeting in Okinawa, Japan. New methodology might also be necessary for the future GMOs in which modifications may be far more complex, and for which current methods may not be able to determine that the GM food is as safe as its counterpart(s). Based on the advances of the genomics area, the starting point is that certain outcomes may assist to address the aforementioned scientific challenges [31]. The 'omics' technologies such as genomics, proteomics and metabolomics provide a 'global' overview of gene expression, protein complement and chemical composition within the crop, be it GM, organic or non-GM. The basics are in place for yielding novel tools that are of generic nature and, therefore, may prove to be valuable for conventional or organic breeding as well. The 'omics' techniques aim to be unbiased with regard to the choice of analytes to be profiled, whether they are genes, proteins or metabolites. Although these methodologies are still in their infancy, they are rapidly developing. There are some early examples of this approach being taken with microarrays and metabolomics. Less appears to have been reported in proteomics. Obviously, this is on account of experimental problems involved in matching up the data between different samples.

Current 'omics' approaches are based on a comparison of the GM-crop with selected counterparts in self contained experiments, such as greenhouse trials. However, it will be of importance of growing trial plants on separate occasions or in different environments. Otherwise what are really 'environmental' effects may be mistaken for effects of the genetic modification. The provision of sufficient and adequate controls guards against this bias. Related to this, there is an urgent need of harmonised protocols for field trials that are specified and documented with respect to:

i)   sufficient number of generations;
ii)   number of locations and growing seasons;
iii)   geographical spreading and replicates;
iv)   statistical models for analysis and confidence intervals, and;
v)   baseline used for consideration of natural variations.

Genomics

This approach provides comprehensive 'snapshots' of the plant cell, tissue or organ at the levels of messenger RNA (the expressed genes or *transcriptome*). The study of gene expression using microarray technology is based on hybridisation of mRNA to a high-density array of immobilised target sequences, each corresponding to a specific gene [32]. MRNAs from samples to be analysed are labelled by incorporation of a fluorescent dye and subsequently hybridised to the array. The fluorescence at each spot on the array is a quantitative measure corresponding to the expression level of the particular gene. Many factors such as maturity stage, and post-harvest effects must be first determined. For instance, green tomatoes differ in their biochemical composition upon ripening from red-ripe fruits [33]. Since hybridisation experiments result in thousands of data points, software packages that can handle these huge amounts of data are essential. The availability of such software is rapidly improving, making it easier to compare large sets of data in order to find similarities and differences in gene expression profiles that may be relevant and can be related to an effect due to genetic modification.

Proteomics

In essence, proteomics is an amalgam of three technologies; high-resolution 2-dimensional electrophoresis (2-DE) to separate the proteins present in a tissue, image analysis to aid comparisons of separations, and mass spectrometry (MS) to determine the identity of the proteins of interest. Correlation between mRNA expression and protein levels is generally poor, as rates of degradation of individual mRNAs and proteins differ. The understanding of the biological complexies in the plant cell can therefore be further expanded by exploiting proteomics, a technique which analyses many proteins simultaneously (i.e. snapshot) and which will contribute to the understanding of gene function. Detection of changes in proteins will also improve the ability to improve the risk assessment by: i) indicating changes in allergenic and or toxic proteins and ii) suggesting changes in the pattern of some metabolites, which would not otherwise be revealed. Antibodies increase the possibilities for detection and identification of proteins from the proteome immunoblot. This may be particularly valuable for the detection of food/pollen allergens by using patients' immune sera.

Metabolomics

The entire collection of metabolites in the cell is called the *metabolome* and the science of measuring it: metabolomics. It is a multi-compositional analysis of the biologically active compounds in plant cells i.e. nutrients, antinutritional factors, toxicants and of other relevant compounds of the metabolome. This analysis may indicate whether intended and or unintended effects have taken place as a result of genetic modification [34]. The four most important techniques that have emerged are various combinations of gas chromatography (GC), high performance liquid chromatography (HPLC), mass spectrometry (MS) and nuclear magnetic resonance (NMR). These methods are capable of detecting, resolving and quantifying a wide range of compounds in a single sample of the plant tissue [36]. With metabolomics it may always be easier to find *some* differences between two sets of samples than to prove conclusively that there is *no* difference. However, as long as the differences are well defined, their importance can be assessed and any uncertainty reduced.

10. Conclusions and Recommendations

The risk assessment of GM crops should focus primarily on the intended novel traits (target gene(s) and derived metabolic product(s)). However, due to the insertional mutagenesis process unintended effects may occur in both GM and non-GM crops, but GM crops are better characterised and it should be bear in mind that risk assessments are not required for non-GM plants. Application of a targeted analysis has its great value and has resulted in a healthy and relative safe food package. The data generated by 'omics' techniques have a potential to increase the knowledge of plant physiology and metabolic networks [35], and eventually, will improve the targeted analyses by discovering additional key nutrients, non-nutrients and toxicants. This scientific challenge provides great advantages for all types of breeding programmes, independently whether they have their roots in traditional, gen-biotech or organic farming. Above all, the aim is to decrease the uncertainties.

Enormous quantities of data can be generated from these 'holistic' methodologies. But do we really reduce the uncertainties in the hazard identification and risk assessment by collecting these huge numbers of data? With regard to their stage of maturity and validation, the subsequent interpretation of the data sets is at present rate limiting. Moreover, 'omics' technologies are not, and may never be, comprehensive. To date, there is also a lack of reports on which to determine the useful contribution of these innovative techniques to the GM-crop risk assessment. Therefore, the 'omics' techniques should not at present be an official requirement. Contrarily, a targeted analysis should still be the leading principle when assessing the substantial equivalence of GM-bred crops. In case of new GM-plant lines with no appropriate comparator or a history of (safe) use, application of the new 'omics' techniques is however of great value for characterisation of their biochemical composition and functions. With respect to this, the improvement of their comprehensiveness of coverage will be one of the major scientific challenges. Thereto, informatic tools need to be developed to extract the most relevant information from the raw data sets.

The development of publicly available databases of crop composition and profiles is an absolute requirement in order to determine the natural variation of compounds within and between given plant species. As information is gathered on natural variation, a (expanding) benchmark on which to compare new crops could be envisaged. These databases would greatly aid the robustness of the targeted analytical approaches. Finally, a major drawback would be the lack of adequate toxicity databases to interpret the safety significance of plant constituents with unknown identity and or function. If not available at the end, the numerous significant differences of unknown structure and or function may lead to consider more extensive safety tests, for example, using laboratory animals.

Acknowledgements

I thank all my colleagues and co-workers who were involved in the EU-funded research projects AGRF-CT90-0039, AIR3-CT94-2311, QLK1-1999-00765 (GMOCARE) and QLK1-1999-01182 (ENTRANSFOOD) and who contributed to the results summarised: Arlette Reynaerts and Barend Verachtert (PGS), Gerrit Alink (WUR), Marcella Pensa (SME Richerche), Howard Davies and Louise Shepherd (SCRI), Jim McNicol (BIOSS), Peter Bramley and Paul Fraser (RHUL), Geert Angenon and Eric Dewaele (VUB), Jan Pedersen and Stine Boerng-Metzdorff (VFA), Friedrich Altmann and Daniel Kolarich (BOKU), Sirpa Kärenlampi, Harri Kokko and Satu Lehesranta (UKU), Ian Colquhoun, Adrian Parr, Marianne Defernez and Yvonne Gunning (IFR), Jean-Jacques Leguay, Michel Péan and Richard Bligny (CEA), Jean-Michel Wal and Gilles Clément (INRA-CEA), Esther Kok, Ad Peijnenburg, Arjen Lommen and Harry Kuiper (RIKILT)

References

(1) James C (2001) Global Review of Commercialized Transgenic Crops: 2001. Ithaca, *International Service for the Acquisition of Agri-biotech Applications*.

(2) EU (1997) 97/618/EC: Commission Recommendation of 29 July 1997 concerning the scientific aspects and the presentation of information necessary to support applications for the placing on the market of novel foods and novel food ingredients and the preparation of initial assessment reports under Regulation (EC) No 258/97 of the European Parliament and of the Council. *Official Journal of the European Communities*, L 253:1-36.

(3) OECD (1993) Safety Evaluation of Foods Derived by Modern Biotechnology: Concepts and Principles. OECD, Paris.

(4) OECD (2001) Consensus documents, OECD Inter-Agency Network for Safety in Biotechnology. OECD, Paris. http://www.oecd.org/oecd/pages/home/displaygeneral /0,33380,EN-document-530-nodirectorate-no-27-24778-32,00.html

(5) WHO (1995) Application of the Principles of Substantial Equivalence to the Safety Evaluation of Foods or Food Components from Plants Derived by Modern Biotechnology. Report of a WHO Workshop. World Health Organization, Geneva.

(6) WHO/FAO (2000) Safety Aspects of Genetically Modified Foods of Plant Origin. Report of a Joint FAO/WHO Expert Consultation on Foods Derived from Biotechnology, 29 May-2 June, 2000.

(7) Kuiper HA, Noteborn HPJM, Peijenburg AACM (1999) Adequacy of methods for testing the safety of genetically modified foods *The Lancet* 354(9187):1315-1316.

(8) Noteborn HPJM (2001) Tackling food safety concerns over GMOs. In: Kessler C, Economidis I (Eds.) EC-sponsored research on safety of genetically modified organisms. European Commission Publications, Luxembourg, 107-109.

(9) SCP (2002) Opinion of the scientific committee on plants on genetically modified high amylopectin potatoes notified by Amylogene HB (notification C/SE/96/3501). Document SCP/GMO/165-final adopted on 18 July 2002.

(10) Noteborn HPJM, Bienenmann-Ploum ME, van den Berg JHJ, Alink GM, Zolla L, Reynaerts A, Pensa M, Kuiper HA (1995) Safety Assessment of the *Bacillus thuringiensis* Insecticidal Crystal Protein CRYIA(b) Expressed in Transgenic Tomatoes. In: Engel K-H, Takeoka GR, Teranishi R. (Eds.) Genetically Modified Foods - Safety Aspects, *ACS Symposium Series* 605:134-147.

(11) Kuiper HA, Noteborn HPJM, Kok EJ, Kleter GA (2002) Safety aspects of novel foods. *Food Research International* 35:267-271.

(12) Ye X, Al Babili S, Kloeti A, Zhang J, Lucca P, Beyer P, Potrykus I (2000) Engineering the provitamin A (beta-carotene) biosynthetic pathway into (carotenoid-free) rice endosperm *Science* 287:303-305.

(13) Fraser PD, Romer S, Shipton CA, Mills PB, Kiano JW, Misawa N, Drake RG, Schuch W, Bramley PM (2002) Evaluation of transgenic tomato plants expressing an additional phytoene synthase in a fruit-specific manner. *Proc Natl Acad Sci USA* 99:1092-1097.

(14) Bovy A, de Vos R, Kemper M, Schijlen E, Almenar Pertejo M, Muir S, Collins G, Robinson S, Verhoeyen M, Hughes S, Santos-Buelga C, van Tunen A. (2002) High-flavonol tomatoes resulting from heterologous expression of the maize transcription factor genes *LC* and *C1*. *Plant Cell* 14 :2509-2526.

(15) OECD (1998) Toxicological and Nutritional Testing of Novel Foods. Report of a OECD Workshop. OECD, Paris.

(16) Codex (2000) Consideration of proposed draft guideline for the conduct of safety assessment of foods derived from plants obtained through modern biotechnology *at step 4*. Codex Alimentarius Commission CX/FBT 01/5.

(17) Gorbunova V, Levy AA (1999) How plants make ends meet: DNA double-strand break repair. *Trends in Plant Science* 4:263-269.

(18) Britt AB (1996) DNA damage and repair in plants. *Annu Rev Plant Physiol Plant Mol Biol* 47 :75-100.

(19) Schnable PS, Hsia A-P, Nikolau BJ (1998) Genetic recombination in plants. *Current Opinion in Plant Biology* 1:123-129.

(20) Hansen G, Wright MS (1999) Recent advances in the transformation of plants. *Trends in Plant Science* 4:226-231.

(21) Kumar S, Fladung M (2000) Transgene repeats in aspen: molecular characterisation suggests simultaneous integration of independent T-DNAs into receptive hotspots in the host genome. *Mol Gen Genet* 264: 20-28.

(22) Beier RC (1990) Natural pesticides and bioactive components in food. *Rev Environm Contamin Tox* 113: 47-137.

(23) Harvey MH, McMillan M, Morgan MR, Chan HW (1985) Solanidine is present in sera of healthy individuals and in amounts dependent on their dietary potato consumption. *Human Toxicol* 4 :187-194.

(24) Coulston F, Kolbye AC (1990) Biotechnologies and food: assuring the safety of foods produced by genetic modification. *Reg Tox Pharm* 12:S1-S196.

(25) Kuiper HA, Kleter GA, Noteborn HPJM, Kok EJ (2001) Assessment of the food safety issues related to genetically modified foods. *The Plant Journal* 27(5):1-28.

(26) Engel KH, Gerstner G, Ross A (1998) Investigation of glycoalkaloids in potatoes as example for the principle of substantial equivalence. In: *Novel Food Regulation in the EU - Integrity of the Process of Safety Evaluation.* Berlin: Federal Institute of Consumer Health Protection and Veterinary Medicine, 197-209.

(27) Gertz JM, Vencill WK, Hill NS (1999) Tolerance of transgenic soybean (Glycine max) to heat stress. Proc International Conference, Brighton, UK, 15-18 November 1999, Vol. 3: 835-840.

(28) GMOCARE (1999) New Methodologies for Assessing the Potential of Unintended Effects in Genetically Modified Food Crops. EU 5FP RTD-project contract No. QLK1-1999-00765; www.entransfood.com/RTDprojects/GMOCARE/default.htm.

(29) Souci SW, Fachmann W, Kraut H (2000) Food Composition and Nutrition Tables. Medpharm GmbH Scientific Publishers, Stuttgart 2000: 1182 pp.

(30) Peijnenburg AACM, Noteborn HPJM, Kuiper HA (2002) Biomarkers for evaluating the safety of genetically modified foods. In: Trull AK, Demers LM, Holt DW, Johnston A, Tredger JM, Price CP (Eds.) Biomarkers of Disease: an Evidence-based Approach. Cambridge University Press, Cambridge, 313-321.

(31) Noteborn HPJM, Peijnenburg AACM, Zeleny R (2002). Strategies for analysing unintended effects in transgenic food crops. In: Atherton KT (Ed) The Gentically Modified Crops: Assessing Safety. Taylor & Francis, London–New York 74-93.

(32) Brown, P.O. and Botstein, D. 1999. Exploring the new world of the genome with DNA microarrays. *Nature Genetics* 21:S33-S37.

(33) Kok EJ, van der Wal-Winnubst ENW, Van Hoef AMA, Keijer J (2001). Differential display of mRNA. *Molecular Microbial Ecology Manual* 8.1.1.:1-10.

(34) Lommen A, Weseman JM, Smith GO, Noteborn HPJM (1998) On the detection of environmental effects on complex matrices combining off-line liquid chromatography and H-NMR. *Biodegradation* 9:513-525.

(35) Noteborn HPJM, Lommen A, van der Jagt RC, Weseman JM (2000) Chemical fingerprinting for the evaluation of unintended secondary metabolic changes in transgenic food crops. *J Biotechnol.* 77:103-114.

(36) Fiehn O (2001) Combining genomics, metabolome analysis, and biochemical modeling to understand metabolic networks. *Comp Funct Genom* 2:155-168.

RISK ANALYSIS OF SOIL-PLANT HORIZONTAL GENE TRANSFER

Enzo Gallori

Department of Animal Biology and Genetics, University of Florence.
Via Romana 17. 50125 Florence. Italy

1. Introduction

The introduction to modern agricultural practice of genetically modified plants (GMPs) has raised a series of questions and concerns that have been debated within the political and scientific communities in recent years (20).[1] Although biological products could offer great potential benefits to agriculture, such as the reduced use of pesticides and fertilizers, there are many uncertainties about the risks of the introduction of transgenic plants into the open environment and the ecological impact of engineered genes. This is based on the hypothesis that if the genetic alteration is transferred to other organisms, in particular microbial recipients, it could be disseminated into the natural habitat, with unpredictable consequences (10).

Studies of the ecological impact of engineered genes have failed to provide a clear consensus about whether GMPs represent an actual risk to the environment, since the data are too few and contrasting. These uncertainties reflect our limited knowledge of the microbial ecology of natural habitats, in particular of agricultural soil which contains and nourishes more than a billion cells per mg of earth (19). Although there has been extensive research on the effects of adding or removing species in communities of higher plants and animals, little information is available for microbial communities. Therefore, an accurate assessment of the risks of GMPs requires an increase of studies on the fate of transgenes after their release into the wild and their influence on natural bacterial communities.

The purpose of this paper is to discuss the possibility of horizontal gene transfer (HGT) from plants to soil microorganisms and the mediating effects of soil mineral components, mainly clay particles, on the transfer processes.

2. Mechanisms of Horizontal Gene Transfer between Bacteria

Horizontal transfer of genetic information between unrelated bacteria has been widely demonstrated in laboratory conditions and in natural systems (2, 21). Recent analyses of the composition of bacterial genomes show that considerable portions of bacterial chromosomes consist of exogenous DNA (about 20% of the genome in *Synechocystis* and 15% in *Escherichia coli*), indicating that the transfer of chromosomal DNA fragments between bacterial species is an important mechanism in their evolution (4). Other studies have shown that genes have been transferred from plants to bacteria during evolution (3).

As plants do not have an identified mechanism of host-range gene transfer, except for pollen hybridization with related species, the possibilities of HGT from plants to bacteria can be explored by considering what is known about the mechanisms of HGT between bacteria.

---

[1] The number(s) between parenthesis indicate the specific literature reference as listed at the end of this paper.

There are three major mechanisms of bacterial genetic exchange: **Conjugation**, **Transformation** and **Transduction** (Fig.1).

Conjugation (Cg) is a process of genetic exchange between bacteria that requires contact between the two cells. Cg was discovered in *Escherichia. coli* by Lederberg and Tatum in 1946, and has been demonstrated in many other gram-positive and gram-negative bacteria. It was the first mechanism of HGT to be studied for genetically modified micro-organisms. It requires the cytoplasmic presence of a circular DNA molecule, a plasmid, which codes for the transfer functions, although both plasmid and chromosomal genes can be transferred. In addition to specific functions for DNA transfer, plasmids codify for many other activities, such as resistance to antibiotics and heavy metals, which can be transferred between bacteria that are very distantly related. Moreover, these broad-host range (BHR) plasmids can mediate the exchange of genetic information even between prokaryotic and eukaryotic cells.

The best known example is the transfer of DNA from *Agrobacterium tumefaciens* to certain plants. *A. tumefaciens* causes a well-known plant disease which consists in the formation of tumors (22). Tumoral transformation is due to incorporation into the host plant's genome of a portion of DNA of plasmid Ti ("Tumor inducing") in the bacterium. The incorporated part (T-DNA) contains genes which control the synthesis of hormones necessary for the formation of tumors; hence the bacterium is called a "natural genetic engineer".

Therefore, conjugation could have played, and could continue to play (plasmids have been detected in plant mitochondria), a significant role in the transfer of genetic information between distant bacterial species, as well as between bacteria and eukaryotic cells. This could explain, at least in part, the numerous discrepancies found in phylogenetic trees.

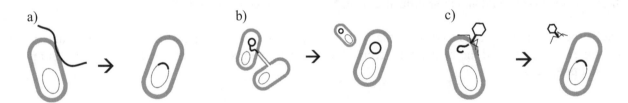

Fig. 1. Schematic representation of the mechanisms of: a) Transformation, b) Conjugation, and c) Transduction.

Transformation (Tf) is the process in which a piece of naked DNA penetrates a recipient bacterial cell, in a physiological state known as "competence", and becomes incorporated into its genome. Tf was the first mechanism of gene transfer studied (6), and its description led to the discovery of DNA as the carrier of genetic information (1). Tf has been demonstrated in more than 40 species of bacteria inhabiting a wide range of environments, such as soil, water and plants.

Although Tf was the first reported instance of bacterial gene transfer in a natural environment (the blood of mice), it was long considered unimportant as a mechanism of HGT in natural habitats. This was due to the two main factors that control the process: the development of competence and the availability of DNA for uptake. Studies carried out in the last ten years have demonstrated that both conditions can occur in soil, and that transformation could be an important mechanism of HGT. The development of competence is highly plausible for two reasons: 1) the presence in soils of protected "microniches" (21), where "competence factors" can accumulate to a concentration that allows the generation on the cells of "transformosomes" for DNA uptake (competence under "external control"); 2) the stressful environmental conditions that usually predominate in soil (limiting concentrations of nutrients or growth factors) can promote competence under "internal control".

Regarding the presence of naked transforming DNA, several studies have shown that DNA in soil can originate from many sources, including lysis of dead cells and the active release of DNA during the "competence" phase of bacterial cells. Observations on the fate of extracellular DNA in natural environments indicate that, despite the presence of various biotic (nucleases) and abiotic (pH, dryness) degrading factors, DNA can persist for a long time as a result of its interactions with soil particles, mainly clay minerals, without losing its ability to transform competent bacterial cells (15). In other words, the interaction (adsorption/binding) of DNA on clay minerals does not prevent its biological activity but rather enhances its persistence in the environment.

Transduction (Td), discovered by Zinder and Lederberg in 1952, is a process that allows the transfer of genes from one bacterium to another through the intervention of virus particles, bacteriophages or "phages". Td has been shown to occur under laboratory conditions in more than 50 bacterial species. However, its importance as a mechanism of HGT in natural habitats, and specifically in soil, is not well understood (21). The major restrictions to Td in soil appear to be due to the necessity of sufficient concentrations of bacteria and phages and to the limited host-range infectivity of phages. Nevertheless, high concentrations of bacteria and bacteriophages have been detected in different types of soil, and phages with host ranges that cross species have been reported. Hence, it is quite possible that Td can occur in the soil environment. Moreover, studies in different laboratories in the last two decades have indicated that the survival and persistence of virus particles in soil is greatly affected by the presence of clay minerals. Adsorption of phages on clay particles protects them against inactivation, e.g. by UV radiation and biodegradation, thus enabling them to persist in terrestrial habitats. Bacteriophages adsorbed on clays are also able to transduce bacteria, indicating that adsorption does not eliminate the viral activity (16).

All these observations indicate that Td could be an important method of HGT in soil.

3. Possibilities of HGT from Transgenic Plants to Soil Bacteria

Recent studies in different laboratories throughout the world have demonstrated that exogenous genes can be transferred, albeit at very low frequencies, from various transgenic plants (sugar beet, potato, tobacco) to soil bacteria (*Acinetobacter* sp., *Pseudomonas stutzeri*) (5, 11, 17). However, the authors have clearly indicated that many physical and biological barriers or factors can hinder HGT between distantly related organisms, thus rendering the risks of spreading transgenic traits (genes or DNA fragments) from GMPs to soil bacteria negligible.

These barriers mainly involve the processes of release and persistence of plant DNA in the soil environment, the uptake of heterologous plant DNA by competent bacterial cells and the phenotypic expression of plant DNA.

With regard to the availability of DNA, studies in different laboratories and in different experimental conditions (including open field tests) have indicated that transgenic plant DNA can persist in the soil for up to 2-3 years, particularly in soils rich in organic matter and clay particles on which nucleic acid can adsorb (12, 18), without inhibiting its availability to competent bacteria. Recently, we developed a very sensitive method to detect plant DNA deriving from GMPs which contained a segment of DNA homologous to bacterial DNA carrying the genes for resistance to the antibiotic kanamycin ("transgenic cassette"); this method was based on the direct extraction of DNA from the soil and its specific amplification by the PCR technique.

Once the plant DNA is available, it must be taken up and internalized by competent bacterial cells. These processes are genetically and environmentally controlled. Although the ability to develop competence under natural conditions has been demonstrated for only a few soil bacteria, the presence in soils of specific microniches, such as aggregates or rizospheres, could permit bacteria to develop competence (21). This suggests that transformation can occur in soils, even if at very low frequencies because of the numerous steps involved in the entire process (10). In fact, the internalization and maintenance of transgenic plant DNA in bacterial cells requires its incorporation into the bacterial chromosome, or its autonomous replication as a plasmid, and the expression of the genetic traits. The incorporation of exogenous genes into the recipient bacterial genome is strictly dependent on sequence homology between the two DNAs (plant and bacterium), and it seems that the degree of homology is the main factor determining the success of stable incorporation of transgenic DNA in bacteria (10). It is assumed that heterologous DNA that penetrates the bacterial wall is rapidly degraded by the bacterial cell's restriction enzyme system. However, in many cases, transformation is mediated by single-strand DNA and thus is not affected by restriction enzymes (9). Moreover, some bacterial species can take up DNA independently of its sequence (14), and bacterial mutants of genes involved in DNA repair processes show less stringent homology requirements (7). Last but not least, many GMPs have been engineered by the insertion of bacterial DNA, which could lead to increased stabilization of plant DNA in bacteria. This is the case of transgenic plants transformed with so-called "gene cassettes", genetic constructions containing bacterial traits like the genes for resistance to antibiotics (13).

The problem of the phenotypic expression of transgenic DNA in competent bacterial cells also represents a strong barrier to successful HGT from plants to bacteria. However, we cannot exclude *a priori* that gene construction can be expressed in bacteria. Indeed, bacterial resistance to antibiotics is developed by the acquisition of resistance genes from heterologous bacterial sources, i.e. by the transfer of plasmids. As many transgenic plants contain both the genes of specific interest and a gene conferring resistance to an antibiotic (for the detection of transformed cells), there is the possibility that these genes could be taken up by indigenous soil bacteria. Moreover, the recent techniques of introduction of genes to plant chloroplasts ("transplastomic plants") could favour the expression of these genes in bacteria because of the prokaryotic-like nature of the chloroplast compartment (8).

4. Conclusions

The results of studies of HGT from transgenic plants to soil bacteria seem to indicate that successful transfer events are extremely rare and that possible risks associated with such events are negligible. Nevertheless, it must be emphasized that this conclusion is based on a small number of observations and indications reported in the scientific literature. As previously mentioned, the main problem in this field is our limited knowledge of the ecology of microbial communities in natural habitats, specifically in soils; thus we probably cannot predict and correctly quantify the actual occurrence of interkingdom gene transfer. Only increased knowledge of the processes regulating the function of microbial communities in different habitats will allow us to predict the consequences of the introduction of transgenic crops into the environment. Until then, we must proceed on a "case-by-case" basis.

References

(1) Avery O.T., McLeod C.M., and McCary M. (1944) J. Exp. Med., 79, 137-159.

(2) Birge E. (1994) *Bacterial and Bacteriophage Genetics*. Springer, New York.

(3) Doolittle R.F., Feng D.F., Anderson K.L., and Alberro M.R. (1990) J. Mol. Evol., 31, 383-391.

(4) Doolittle R.F. (1999) Science, 284, 2124-2128.

(5) Gebhard F., and Smalla K. (1999) FEMS Microbial. Ecol., 28, 261-272

(6) Griffith F. (1928) J. Hyg., 27, 113-159.

(7) Harris R.S., Langerich S., and Rosenberg S.M. (1994) Science, 264, 258-260.

(8) Kay E., Vogel T.M., Bertolla F., Nalin R., and Simonet P. (2002) Appl. Environ. Microbiol., 68, 3345-3351.

(9) Lorenz M, and Wackernagel W. (1994) Microbiol Rev, 58, 563-602

(10) Nielsen K.M., Bones A.M., Smalla K., and van Elsas J.D. (1998) FEMS Microbiol. Rev., 22, 79-103.

(11) Nielsen, K. M., van Elsas J.D., and Smalla K. (2000) Appl. Environ. Microbiol., 66, 1237-1242.

(12) Paget E., Lebrun M., Freyssinet G, and Simonet P. (1998) Eur. J. Soil Biology, 34, 81-88.

(13) Recchia G.D., and Hall R.M. (1995) Microbiology, 141, 3015-3027.

(14) Stewart G.J. (1989) In: *Gene Transfer in the Environment* (S.B. Levy, and Miller R.V. eds.), 139-164. McGraw-Hill, New York.

(15) Stotzky G., Khanna M., and Gallori E. (1996) In: *Molecular Microbial Ecology* (A.D. Akkermans, J.D. van Elsas, and F.J. de Brujin eds.), 1-28. Kluwer, Dordrecht.

(16) Vettori C., Stotzky G., Yoder M., and Gallori E. (1999) Environ. Microbiol., 1, 251-260.

(17) de Vries J., Heine M., Harms K., and Wackernagel W. (2003) Appl. Environ. Microbiol., 69, 4455-4462.

(18) Widmer F., Seidler R.J., Donegan K.K., and Reed G.L. (1997) Mol. Ecol., 6, 1-7.

(19) Williamson M. (1992) Mol. Ecol., 1, 3-8

(20) Wolfenbarger L.L. and Phifer P.R. (2000) Science, 290, 2088-2093

(21) Yin X. and Stotzky G. (1997) Adv. Appl. Microbiol., 45, 153-212.

(22) Zambrynski P., Tempe J., and Schell J. (1989) Cell, 56, 193-201.

SESSION 5:
THE VARIOUS STAKEHOLDERS' POSITIONS

SUSTAINABLE AGRICULTURAL SYSTEMS AND GMOs.
IS CO-EXISTENCE POSSIBLE?

Alexandra Hozzank

InfoXgen-Working Group Transparent Food.
Königsbrunnerstr. 8, 2202 Enzersfeld, Austria

Abstract

European agriculture is characterised by a small-scale structure and by a high share of organic farmers. Traditional production systems are also preferred in conventional agriculture. The EU Regulation for organic farming no. 2092/91 forbids strictly the use of GMOs. Farmers, ecologists as well as consumers in Europe have quite a critical attitude towards genetic techniques, as they are seen as not necessary and dangerous in certain areas. In Europe, except Spain, there are no GMOs cultivated commercially. There are no long-term studies about the effects of GMOs on ecosystems. In addition, the problems of co-existence, measures to avoid contamination, liability, costs and others, have not been solved yet. On the basis of the studies published up to now, a co-existence of organic, GMO-free and GMO agriculture is not possible. Therefore, in spite of an enormous pressure on the part of the WTO and the USA, the attitude of Europe is, to prevent the cultivation of GMOs until the issues of co-existence are solved.

1. Introduction

In 1996 genetically modified plants were first commercially planted worldwide, since then the cultivation of GM crops increased rapidly.

About 51 % of the soja, 9 % of the of maize and 12 % of the oil seed rape production deriving from GMO plants, is grown mainly in USA, Argentina, Canada, China and recently in Brazil too (soja production). Spain is the only country in Europe, where GMOs (Bt-Maize, about 25.000 ha[1]) are cultivated.

In Europe farmers as well as consumers are quite sceptical towards genetic modified crops and food, the consequence of this attitude was the creation of a control system, which should prevent the authorization of GM crops up to the moment the environmental and health risks would have been clarified completely.

In June 1999 a de-facto moratorium on GM products was established. Its main task was to control the authorization and placing on the market of GM crops in Europe. The moratorium was meant to be maintained up to the revision of the Regulation 90/220[2], which mainly regulates the setting free of GMOs. In February 2001 the revised Regulation 2001/18 was adopted by the European Parliament and entered into force in October 2002. The United States joined by other states called for a removal of the moratorium within the WTO and therefore for free export possibilities of GM crops to Europe.

As far as it concerns Europe the new legislation does not regulate the GMOs problematic, also because of the completely different agricultural situation. Europe has a high share of organic farmers, as well as a small structured agriculture.

---

[1] Hacres.

[2] For an overview of the EU legislation quoted in the text please see paragraph 7. Legislative Framework of this paper.

The EU Regulation 2092/91 which regulates organic farming and processing systems, prohibits the use of GMOs, therefore, new EU legislation regulating the co-existence of organic, GMOs free and GMOs cultivation had become necessary.

In July 2003 the European Commission issued guidelines for co-existence, which are based mainly on the voluntary arrangements between neighbouring farmers, and try to solve the problem on national level. Therefore, the demand rose to keep up the Moratorium, up to the moment the problem of co-existence has been solved.

2. Problems of Co-Existence

2.1. Routes of contamination[3]

2.1.1. Seeds and Seedlings

The purity of purchased seeds influences remarkably the contamination level of the harvest. Although certified seed production has to meet high quality standards, it can happen that also certified seed is mixed during storage or transportation. Up to now there is very limited GM production in Europe. Therefore, the danger of seed contamination is not that high. However, the case in Upper Austria, where Maize seed was mixed with a small amount of GM seed from the US, showed that farmers always have to face the contamination problem.

2.1.2. Transfer of Modified DNA

Pollen Transport and Fertilisation
The pollen dispersal determines to a large extent the distances of vertical gene transfer. The main pollinators are wind and insects, as well as birds and mammals. For some plants water as well can play an important role in pollination. For estimation of distances climate and topography has to be taken into consideration too. For the species where GMOs are cultivated commercially the pollination is limited to wind and insects.

Diaspore Banks
The transportation of seeds has to be taken into account as well when trying to estimate gene flows. Oilseed rape can persist in soil for many years (Lutman 1993, Sauermann, 1993, Schlink 1994 and this can lead to severe volunteer problems.

Volunteer rape is a common and widespread weed in cereal rotations and field margins, roadsides and soil dumps. Therefore, volunteers can also act as a reservoir of transgenes. Feral rape populations can persist around agricultural land for up to 10 years.

Transfer of Genes from Cultivated to Wild Species
Some plant species, e.g. oilseed rape, have the capacity to build up independent populations outside the cultivated area. After feral plants have become established in various ecosystems, management measures are only sufficient to a limited extent.

---

[3] Contamination can be seen in this context. Conventional and/or organic crops and products can be made unsuitable for the market if mixed with GM components above certain thresholds.

Any gene that leads to an increase in fitness, such as is potentially presented by virus or insect resistance, is more likely to persist.

Horizontal Transfer
There are various uncertainties about the horizontal gene transfer from plants to soil microorganisms. Studies showed that there is an accumulation of toxin, deriving from
Bt-Maize in soil, retaining its anti-lepidopteran activity for at least 180 days (Saxena et al., 2002).

2.1.3. Harvesting and Processing

Seeds can be inadvertently moved from field to field in machinery, as well as spillage occurring along transport corridors. Oil seed rape ruderal populations for example were found along railway tracks in Lower Austria (Pascher et al., 2000).

Contamination of other crops can occur through gene flow from volunteers or from volunteers being harvested along with the crop. Potential sources for contamination are seed drills, cultivation equipment, harvesting machinery, transporting vehicles, storage, cleaning, drying facilities and packing.

In the case of processing of GM and non-GM products in the same installations there is a very high risk of contamination, since complete cleaning is not possible for dusty goods. This is also one of the main problems in feed production. Ingredients, additives and processing agents may consist of or contain GMOs as well.

2.2. Possible Measures to Avoid Contamination

2.2.1. Measures to Avoid Pollen Transfer

Security Distances
Oil seed rape presents a high risk for cross-pollination between source and recipient fields. Pollen dispersal has been recorded at up to 4km by honeybees (Ramsey at al, 1999), and to 3km by the air flow (Thompson at al., 1995).

Sugar beet presents a medium to high risk for cross-pollination both with other stands and with wild relatives. In areas producing sugar beet, seed flowering is a necessity and here the risk of cross-pollination increases accordingly. The pollen produced can be spread extensively on the airflow up to 800m (Jensen et al., 1941).

Maize presents a medium to high level of risk for cross-pollination with other maize crops as the pollen can spread on the airflow. Pollen distribution, as determined by outcrossing between different maize varieties, has been recorded at up to 800m (Jones et al. 1950). Maize also presents a medium to high risk for the inclusion of pollen into honey.

Wheat, Potato and Soja can be described as low risk for contamination from genetically modified varieties.

Much of the research published relies on small field trials (including GM trials). Evidence indicates that the extent of gene flow between GM and non-GM fields, and between GM and feral populations depends mainly on the scale of pollen release and dispersal, and on the distances between source and recipient populations.

The potential impact (including cross pollination and inclusion in honey) of pollen from GM crops increases notably with the size and number of fields planted.

Hedges and Similar Barriers
The few studies on this issue show the uncertainty associated with the establishment of border rows or barriers to reduce gene flow (Ingram 2000, Morris et al 1994). A lot of research in this area has still to be done.

Genetic Engineering Approaches
Genetic engineering methods are also seen to provide pollen dispersal. According to Eastham and Sweet (2002) following techniques are worth considering with regard to GMOs:
- Apomixes: production of seeds without fertilisation
- Kleistogamie: self-pollination in closed blossom
- Prevention of flowering by subsequent control of flowering through application of chemical elicitors
- Male sterility: prevention of flowering and development of pollen
- Plastid transformation: Plastid DNA is transformed instead of DNA of the nucleus. In many cases, this allows to prevent transgenic pollen, as plastids are maternally inherited in most macrophytes.
- Sterility of seeds: prevention of germination of seeds

There are no indications up to what extent these possibilities are merely theoretical in nature and bring yield declines with them.

GMO Free Areas
The establishment of GMO free areas is one of the efficient measures to avoid contamination. The determination of protection goals is here of central importance. Some of these evaluation criteria could be:
- protected areas for preservation of biodiversity (e.g. Natura 2000)
- areas for organic farming
- areas of the enhanced in-situ (on-farm) preservation of plant-genetic resources under GMO-free conditions (Art.8 of the CBD)
- development or "transition" areas for sustainable agricultural development
- mountain areas, whose ecological sensitivity merits special consideration following Agenda 21,Capt.13

3. Thresholds

Each farmer should have the possibility to choose the agricultural systems, most appropriate to him. This freedom of choice needs a legal basis, such as strict threshold values. Organic certification works on the basis of the traceability principle. Testing of the end product is a complementary tool to confirm that the control system is sufficient, but can never be the main goal. Therefore, threshold levels can only indicate the maximum tolerance for exceptional and unforeseeable contamination events but not for permanent levels of contamination.

The following thresholds were established for Austria:

Threshold	Valid for	Reference
0,1	Seed of the following species: • Swede (Brassica napus L. var. napobrassica) • Maize (Zea mays) • Oilseed rape (Brassica napus) • Turnip rape (Brassica rapa) • Soybean (Glycine max) • Turnip (Brassica L. var. rapa) • Tomato (Lycopersicon lycopersicum) as processing varieties • Chicory (Cichorium intybus L.)	BMLFUW 2001w
0,1	Threshold values for exceptional and unforeseeable contamination with GMOs and derivates for products deriving from organic agriculture: for food ingredients and food processing aids, feed ingredients and feed processing aids, fertilisers and soil improvers	BMSG 2001

Recently the EU Regulation 1829/2003 on genetically modified food and feed had been published. Food and feed that contains more than 0,9% GMOs or derivates related to the different ingredients, has to be labelled, irrespective of the detectability of transgenic DNA or protein. Until now there are no threshold values for organic products established by the EU. A threshold value of 0,9% is not compatible with organic standards. The threshold should be based on the detection limit, which is now 0,1%, also in order to meet consumers interpretation of organic or GMOs free products. The planned fixing of threshold values for seed on EU level had been recently postponed because of inconsistency between member states. Austria as well as Italy stands up for values on the detection limit. In the case of commercial cultivation of GMOs a thresholds of 0,1% seems to be extremely difficult to achieve (Bock et al, 2002).

4. Liability

If adventitious presence of GM crops in non-GM crops occurs above a set threshold a reduction in income could be expected. Organic farmers could also loose their subsidies. Therefore, a clear liability program has to be in place before GM cultivation starts.

In addition to the strict liability regime, IFOAM[1] proposes also the establishment of a compensation fund, paid for by the GMOs industry. The fund would use the contributions from all who benefit financially from GMOs to compensate those who suffer financial loss through genetic contamination. It would also cover any environmental damage (IFOAM, 2003)

5. Agricultural Structure in Austria

Austria is characterised by small structured agriculture, the average farm size is between 10 and 20 ha and by a very high percentage of organic farmers.

---

[1] IFOAM stands for: International Federation of Organic Agriculture Movements (www.ifoam.org).

The organic movement in Austria originated in 1980, when 200 farmers decided to cultivate their farms according to organic methods. The development had its largest growth up to the end of the nineties; in 1999 it reached its peak with 19,733 farms.

The most important factors for this enormous increase have been:

- the early addition of guidelines for organic crop production and animal husbandry to the Austrian Codex Alimentarius;
- the government support for organic farms during and after conversion through compensatory payments;
- favourable market conditions through the entrance of supermarket chains and
- an active policy of the organic associations (mainly BIO ERNTE AUSTRIA, with a share of nearly 65% of all organic farmers).

In 2001 an "Action Plan for Organic Agriculture in Austria" was developed. This plan is scheduled to increase the organically treated area to 50% until 2006 (BMLFUW, 2003b).

In 2002 17.891 farmers cultivated 12% of the agriculturally used area (BMLFUW, 2003a).

Moreover, a high percentage of Austria belongs to ecological sensible areas like the Alps or Natura 2000 areas.

Because of the characteristics mentioned above and also because the clear attitude of consumers and farmers against genetic techniques in food and feed production, Austria tried to establish (national or regional) regulation to forbid the cultivation of GMOs.

The national law from Upper Austria, creating a GMO free zone Upper Austria, was rejected by the European Commission following the assessment of the European Food Safety Agency (see EU press release IP/03/1194). The reason for this rejection was the lack of scientific evidence, confirming that Upper Austria has extraordinary topographic and ecological characteristics.

6. Conclusion and Recommendations

Based on the European Commission recommendations on co-existence there is a strong need to regulate co-existence of organic, GMO-free and GMO farming systems on all levels, from farm level to management at the neighbourhood level and measures with region-wide dimension. All effective and necessary measures to avoid contaminations are extremely expensive and need time consuming planning phases. Therefore, questions concerning liability must be solved completely before GM commercial cultivation is possible in Europe.

If organic farming as well as certified GMO-free production are going to continue and if consumers expectations and preferences in the long term have to be met, there is a strong need for additional protective measures for the organic production system. Particularly, defined areas are required which can be used to build up and maintain a separate, GM-free branch of seed breeding and propagation.

7. Legislative framework

EU Legislation:

- Directive 2001/18/EC of the European Parliament and of the Council of 12 March 2001 on the deliberate release into the environment of genetically modified organisms and repealing Council Directive 90/220/EEC 23 April 2003 (OJ L117,08/03/1990 p. 15) - Commission Declaration

- Regulation (EC) No 1829/2003 of the European Parliament and of the Council of 22 September 2003 (OJ L268 18/10/2003 p.1) on genetically modified food and feed (Text with EEA relevance)

- Regulation (EC) No 1830/2003 of the European Parliament and of the Council of 22 September 2003 (OJ L268 18/10/2003 p.24) concerning the traceability and labelling of genetically modified organisms and the traceability of food and feed products produced from genetically modified organisms and amending Directive 2001/18/EC

- Council Regulation (EEC) No 2092/91 of 24 June 1991 on organic production of agricultural products and indications referring thereto on agricultural products and foodstuffs

- Commission Recommendation 2003/556/EC of 23July 2003 (OJ L189 29/07/2003 p.36) on the guidelines for the development of national strategies and best practices to ensure the co-existence of genetically modified crops with conventional and organic farming.

- Commission Regulation (EC) No 49/2000 of 10 January 2000 amending Council Regulation (EC) No 1139/98 concerning the compulsory indication on the labelling of certain foodstuffs produced from genetically modified organisms of particulars other than those provided for in Directive 79/112/EEC

- Corrigendum to Commission Regulation (EC) No 50/2000 of 10 January 2000 (OJ L 6 11/1/2000 p.) on the labelling of foodstuffs and food ingredients containing additives and flavourings that have been genetically modified or have been produced from genetically modified organisms

Austrian Regulations:

- BMLFUW, 2001: Verordnung 478 des Bundesministeriums für Land- und Forstwirtschaft, Umwelt und Wasserwirtschaft über die Verunreinigung von Saatgut mit gentechnisch veränderten Organismen und die Kennzeichnung von GVO-Sorten und Saatgut von GVO-Sorten (Saatgut-Gentechnik-Verordnung).BGBl II/478/2001.

- BMSG, 2001: Beschluss betreffend "Festlegung von Schwellenwerten für zufällige, unvermeidbare Verunreinigungen mit gentechnisch veränderten Organismen und deren Derivaten" zur Verordnung (EG) Nr. 2092/91, Biologische Landwirtschaft. GZ.32.046/72-IX/B/1b/01.

Private Standards:

- IFOAM Position Paper, IFOAM EU Group, 2003: Position Paper: Co-existence between GM and non-GM crops, Necessary anti-contamination and liability measures. http://www.ifoam.org/pospap/ge_position_0205.html

References

(1) BMLFUW, 2001: Verordnung 478 des Bundesministeriums für Land- und Forstwirtschaft, Umwelt und Wasserwirtschaft über die Verunreinigung von Saatgut mit gentechnisch veränderten Organismen und die Kennzeichnung von GVO-Sorten und Saatgut von GVO-Sorten (Saatgut-Gentechnik-Verordnung).BGBl II/478/2001.

(2) BMLFUW, 2003a: Grüner Bericht, Wien (in press).

(3) BMLFUW, 2003b: Aktionsprogramm. Biologische Landwirtschaft 2003-2004.

(4) BMSG, 2001: Beschluss betreffend "Festlegung von Schwellenwerten für zufällige, unvermeidbare Verunreinigungen mit gentechnisch veränderten Organismen und deren Derivaten" zur Verordnung (EG) Nr. 2092/91, Biologische Landwirtschaft. GZ.32.046/72-IX/B/1b/01.

(5) Bock A-K., Lheureux K., Libeau-Dulos M., Nilsagård H. und Rodriguez-cerezo E., 2002: Scenarios for co-existence of genetically modified, conventional and organic crops in European agriculture. Joint Research Centre, European Commission.

(6) Eastham K, Sweet J., 2002: Genetically modified organisms (GMOs): The significance of gene flow through pollen transfer. Report, Environmental issue report No 28, A review and interpretation of published literature and recent/current research from the ESF "Assessing the Impact of GM Plants" (AIGM) programme for the European Science Foundation and the European Environment Agency.

(7) IFOAM EU Group, 2003: Position Paper: Co-existence between GM and non-GM crops, Necessary anti-contamination and liability measures. http://www.ifoam.org/pospap/ge_position_0205.html.

(8) Ingram J, 2000: Report on the separation distances required to ensure cross-pollination is below specific limits in non-seed crops of sugar beet, maize and oilseed rape. National Institute of Agricultural Botany, Cambridge UK. Published by Ministry of Agriculture , Fisheries and Food –UK (MAFF).

(9) Jensen I. And Bogh H. 1941: On conditions influencing the danger of crossing in case of wind pollinated cultivated plants. Tidsskrift for Planteavl, 46, 238-266.

(10) Jones M.D. and Brooks J.S., 1950: Effectiveness of distance and border rows in preventing outcrossing in corn. Oklahoma Agricultural Experimental Station. Bulletin no. T-38.

(11) Lutman P.J.W., 1993: Volunteer crops as weeds, Aspects of Applied Biology 35, pp. 29-36.

(12) Morris W.F., Kareiva P.M. and Raymer, P.L., 1994: DO barren zones and pollen traps reduce genetic escape from transgenic crops? Ecological Applications, 4, 157-165.

(13) Müller W., 2003: Concepts for Coexistence, Final report. September 2003.

(14) Pascher K., Macalka-Kampfer S., Reiner H., 2000: Vegetationsökologische und genetische Grundlagen für die Risikobeurteilung von Freisetzungen von transgenem Raps und Vorschläge für ein Monitoring. Bundesministerium für soziale Sicherheit und Generationen, Forschungsberichte 7/2000.

(15) Ramsey G., Thompson C.E., Neilson S. and Mackay, G.R., 1999: Honeybees as vectors of GM oilseedrape pollen. In: Lutman P.J.W. Gene flow and Agriculture: Relevance of Transgenic Crops. BCPC Symposium Proceedings no. 72.

(16) Sauermann, W., 1993: Raps. 11(2), pp. 82-86.

(17) Saxena D., Flores S. und Stotzkya G., 2002: Bt toxin is released in root exudates from 12 transegenic corn hybrids representing three transformation events. Soil Biology & Biochemistry 34.

(18) Schlink S., 1994: Ökologie der Keimung und Dormanz von Körnerraps (*Brassica napus* L.) und ihre Bedeutung für eine Überdauerung der Samen im Boden. PhD Thesis, University of Göttingen, Germany.

(19) Thompson C.E., Squire G., Mackay G.R., Bradshaw J.E., Crawford J. and Ramsay G., 1999: Regional patterns of gene flow and its consequence for GM oilseed rape. In: Lutman P.J.W. Gene flow and Agriculture: Relevance of Transgenic Crops. BCPC Symposium Proceedings no. 72.

THE PROCESSING SECTOR: INTEGRATION OF ENVIRONMENTAL CONCERNS IN INDUSTRIAL STRATEGIES

Piet W.M. van Dijck[1,] and Philip H. van Lelyveld[2]
DSM Life Sciences.
PO Box 1, 2600 MA, Delft, The Netherlands

Abstract

Six examples of innovations with modern biotechnology are discussed in terms of their effects on industrial economic value creation, environmental performance and on the effects on people's lifestyles and working conditions. The examples span a period of 3 decades. Gradually increasing levies on end-of-pipe pollution and societal pressure for cleaner processes, technology development and the need to compete on a global market for food, feed and pharmaceutical ingredients all four resulted in new and innovative solutions. We conclude that industry can and will indeed respond to environmental concerns by developing cleaner technology and will come with innovative products and processes. However these developments need sufficient time and sometimes support from national and local governments before society in a broader sense can benefit from them.

1. Introduction

Since its founding in 1902 as the Dutch State Mining company, DSM has undergone many changes: from coal mining to base chemicals, from base chemicals to specialty chemicals, from regional to global from state-owned to a publicly owned. But other things remained: our knowledge and experience in technology and manufacturing, our constant search for innovation, and our conviction that it is people and relationships between people that determine the success of the company. Like many organisations DSM has adopted the idea of Triple P – people, planet and profit. For DSM this means achieving an equitable and effective balance in everything we do: running a profitable business, treating people fairly and concern for the environment in sustainable development. Embracing sustainable development means being prepared to change what we do – not just technical actions like reducing pollution from a factory pipe. Where do we stand on biotechnology, responsible marketing, product stewardship, animal testing and community dialogue? We are learning that being sustainable is a balancing act: juggling the demands of business in a competitive market with minimising environmental impact and respecting the interests of the people who make, buy and use our products. Business must be responsible throughout the entire value chain of their products. How we conduct ourselves, how we are organised, how we decide and what we deliver, as well as how much profit we make. And this all begins with principles such as transparency, partnership and good corporate citizenship.

In the field of Life Sciences many of DSM's products are based on fermentation processes with micro-organisms. In the next pages we will give a number of examples to show how this has led to innovative products and processes.
In many of the production processes the producing micro-organism has been modified and improved for overproduction of the desired product by genetic modification (recombinant) techniques (Genetically Modified Micro-organisms or GMMs). However, just to show that rDNA technology is just one of the many technologies our R&D people use in their toolbox to develop innovative products and processes we

will start with two examples based on classical microbiology and/or conventional mutation and selection techniques for improvement of the productivity.

2. Discharge of aquaous wastes from a fermentation plant

In the 70s environmental awareness led to the introduction of a tax for the discharge of aqueous biodegradable wastes, such as the wastes from fermentation processes, into the environment. The costs of this tax for the Delft fermentation plant of Gist-brocades (now DSM) would be so high that it would seriously threaten the economic viability of the site as a production plant for bakers yeast, penicillin and other fermentation products.

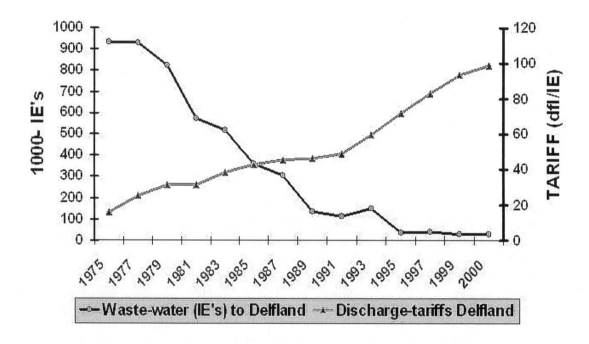

Figure 1.   Discharge of wastewater (in IE's – a measure of the amount of biodegradable material present in the waste) from the DSM production site to the municipal surface waters correlated to the municipal tariffs (in Dutch guilders per IE).

Therefore R&D was asked to develop a reproducible water purification system with controllable sludge properties that is suitable for extremely variable industrial wastewater. Supported by governmental funds, this has led to the development of an innovative sludge-on-carrier technology, fluid bed reactors in which the micro-organisms in sludge are attached to fluidised sand grains as a carrier. This anaerobic water purification process was put in process in the late 70s and reduced the discharge of waste into the environment gradually over time to about 10% of its original value. Some of the reductions are accounted for by increasing the capacity of the wastewater purification plant whereas other reductions are due to innovations in the production processes, e.g. enzymatic processes instead of chemical conversions. The latter eliminated the use of organic solvents, which were very detrimental to the proper functioning of the microbes in the wastewater purification system. In 1993-1995, when the taxes had increased considerably, introduction of a aerobic waste-water treatment facility, coupled to the anaerobic reactors, made it possible to reduce the discharge to about 2% of its original value (figure 1). A win-win situation: minimisation of the discharge of biodegradable organic waste (environmental benefit - planet) and restoring the cost-effective position of the Delft production site (people and profit).

3. Production Improvement Penicillin-G

Most antibiotics used around the world are so-called beta-lactam antibiotics, antibiotics based on a penicillin or a cephalosporin nucleus. The notion that molds may produce anti-bacterial compounds is based on an observation by Alexander Fleming. He observed that when an agar culture of bacterial cells became infected by a mold colony bacterial cells could not grow in the vicinity of the mold colony. This bacteriostatic effect of the Penicillium mold could be attributed to the production and secretion of penicillin by the mold. Some research was done to purify and characterize the compound but it was not until the outbreak of World War II that the issue became important. Due to massive losses on the battlefields as a consequence of bacterial infections in wounded soldiers the Allied forces decided that they needed good antibacterial agents. The Anglo-American secret war project on penicillin had the highest priority next to the development of the atomic bomb and by 1943 the Allied forces produced enough penicillin to start treating wounded soldiers on the battlefields successfully.

The success of this wonder drug also spread clandestinely to Nazi-occupied Europe. Devoid of UK/US information sources on penicillin researchers at the Nederlandsche Gist- and Spiritusfabriek (Dutch Yeast and Spirit Factory)(NG&SF) managed to develop a fermentative production process for penicillin in the last year of the war (Burns and van Dijck, 2001; Burns et al, 2002). Already in 1949 the penicillin production had reached levels sufficient to meet demand in The Netherlands, enabling export The Anti-Infectives Business Group of DSM now is the world's leading producer of bulk penicillin. It soon was apparent that penicillin, or more precisely penicillin-G, was poorly acid-stable and was not effective towards all important pathogenic bacteria. Therefore penicillins were derived from penicillin G by replacing the penicillin G phenylacetyl side chain chemically (and later on enzymatically) by other side chains, which rendered these so-called semi-synthetic penicillins more effective towards other groups of bacteria and made the product resistant to the acidic environment of the stomach. Examples of such semi-synthic penicillins are ampicillin and amoxycillin. So penicillin-G is considered a raw material for semi-synthetic penicillins and as productivity increased penicillin-G turned into a commodity product. The competitive field changed dramatically in the last decades of the 20th century. Whereas in the 70s all main competitors were pharmaceutical companies from Europe and the USA (Hersbach et al., 1984), presently the main competitors are companies from China and India. Competition on costprice therefore is the key-parameter to stay in business. DSM's main production site in Delft, where the costs of both employment and energy are high, can only compete with other producers in lower-cost countries by virtue of a continued investment in R&D programmes to improve the penicillin production level of the producing strain by classical mutation and selection techniques. The constant improvement in penicillin-G productivity in DSM (figure 2) can to a significant extent be attributed to the development of improved strains.

Development in productivity

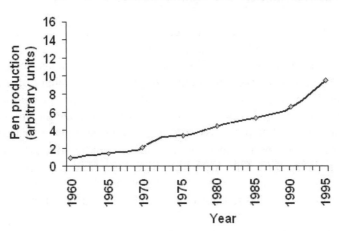

Figure 2.   The development in production of Penicillin-G in DSM.

4. Green Production of Cephalexin

It soon turned out that some bacteria developed resistance towards penicillins and this started a search for related antibiotics. In the early fifties it was found that the fungus *Cephalosporium acremonium* produced cephalosporin-C, also a beta-lactam antibiotic but with a 6-membered ring adjacent to the beta-lactam ring instead of the 5-membered ring structure of penicillin. This opened the opportunity to develop a series of semi synthetic cephalosporins as was done with the penicillins. However the production of cephalosprin by fermentation was hampered by the fact that the cephalosporin-C produced by *Cephalosporium acremonium* is rather water soluble. Whereas penicillin is recovered by a simple extraction and crystallization the recovery of cephalosporin-C from the broth requires an expensive purification route. In the 60s and 70s Gist-brocades developed a chemical process to produce from penicillin-G a common intermediate 7-amino-desacetoxy-cephalosporanic acid (7-ADCA) for the semi-synthetic production of economically important cephalosporins such as cefadroxil, cephradin, and cephalexin (Hersbach et al., 1984).

Since biochemical and genetic investigations had revealed that penicillin and cephalosporin have the first part of the biochemical pathway in common, the use of rDNA technology made it possible to engineer a high producing *Penicillium chrysogenum strain* in such a way that it now started to produce a cephalosporin molecule (Elander, 2003). The produced cephalosporin molecule then can be easily converted by one enzymatic step to 7-ADCA and by another enzymatic step to cephalexin. In this manner a complex traditional chemical process was replaced by a "green" process consisting of a fermented intermediate linked enzymatically with a side chain to the final end product.

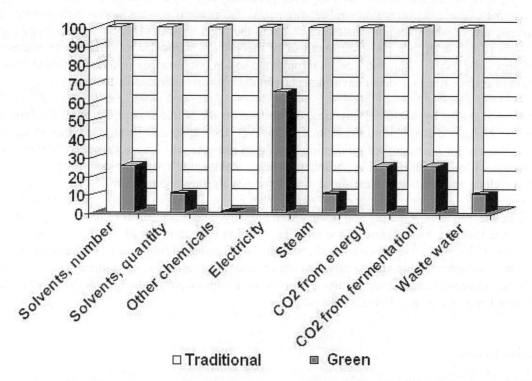

Figure 3. Environmental savings of the Green versus the Traditional manufacturing process of Cephalexin in percentages.

This example of metabolic pathway engineering (Anonymous, 2001) was developed at DSM and was scaled up to commercial production in 2001. The impact of this green process on the environment was calculated by the German Öko Institute (Sijbesma, 2003).

Besides a reduction of the variable costs by 50% the environmental savings also were considerable: 65% in reduction of materials used and 65% savings in energy consumption (Sijbesma, 2003). This once more is an environmental (planet) benefit, coupled to both benefits for profit (the economics of the process) and the people (the workers at the Delft production site).

5. Phytase

Monogastric animals, such as pigs and poultry, are incapable of using organically-bound phosphate, present in vegetable feeds. Due to this inability feed producers have to add inorganic phosphate to the feeds to meet the animals nutritional mineral requirements. Phosphate is present in vegetable material in the form of phytate, in which six phosphates are bound to the sugar inositol. This bound phosphate ends up in the manure and adds to a phospate overload of the soils in particular in areas where intensive farming of pigs and poultry occur, and leaches through the soil into the surface waters. Phosphate causes eutrophiation of surface waters, a major environmental problem in areas of intensive farming.

Phytate can be degraded by microbial phytases, which liberate the bound phosphates from the inositol and makes the phosphate available to the animals. However, the expression level of phytase in micro-organisms is too low for the microbial production process to compete with its traditional counterpart, the supplementation of feeds with inorganic phosphate. The recombinant DNA technology has provided the possibility to clone the phytase gene from a wild-type Aspergillus niger strain and to over-express multiple gene copies in an industrial strain of *Aspergillus niger*. The product marketed as Natuphos is available since the mid 90s for the pig and poultry feed industry. The replacement of inorganic phosphate by Natuphos has resulted in a decrease of the phosphate load in the manure of about 30%. Consequently the use of recombinant DNA technology has significantly contributed to relieve a serious environmental problem in areas of intensive pig and poultry farming or in areas of intensive fish farming (aquaculture). Without the use of recombinant DNA technology an economically viable production method for phytase would not exist today (van Dijck, 1998).

6. Vegetarian Cheese

Rennet, an enzyme mixture from the stomach of suckling calves and other ruminant young animals, is essential in cheese making. In order to inducing curdling, precipitation of the milk protein, rennet has to be added to the milk. The responsible enzyme in the rennet to induce curdling is chymosin. Gist-brocades pioneered to develop a production process of chymosin by over-expressing the gene from the calves stomach by a dairy yeast *Kluyveromyces lactis* (van Dijck, 1988). This chymosin is the same as that isolated from calves stomachs. The product was in 1988 the first product introduced onto the food market produced with a GMO, and this production method has environmental benefits too. The microbial process is far more efficient than the traditional production by extracting the chymosin from calves stomachs. This saves a lot of energy and produces less waste. In addition the availability of calves stomachs is limited. Rennet production for cheese requires 10-20 million calves stomachs a year that have to be transported from all over the world to the few animal rennet producers in the world. Lastly, and this is a peoples benefit, since the chymosin produced by the GMO yeast complies with kosher, halal and vegetarian requirements, these consumer groups can consume cheese without conflicts with their eating practices. Cheese made with chymosin from a GMO yeast can thus be labelled as vegetarian cheese (figure 4).

Figure 4.   A label from a UK cheese made with chymosin from a GMO yeast, carrying the label of the Vegetarian Society.

7. An Innovative Enzyme Production Technology in *Aspergillus Niger*

For DSM the fungus *Aspergillus niger* is of eminent importance; many enzymes for the food and feed industry are produced using A. niger strains. Furthermore its ability to produce citric acid was exploited on an industrial scale firstly in 1919 by Citrique Belge, a company recently acquired by DSM, as part of the Roche Vitamins and Fine Chemicals acquisition. *Aspergillus niger* is generally regarded as a safe and non-pathogenic fungus widely distributed in nature (Schuster et al, 2002). Therefore it has been approved as GRAS by the US FDA for the production of a number of food enzymes. The only safety issue for this fungal species is that a small percentage of strains have the ability to produce Ochratoxin A, a mycotoxin. All of the DSM GMO A. niger enzyme production strains lack this capacity, and are based on the industrial strain overproducing the enzyme glucoamylase. This enzyme was amongst the enzymes which were affirmed GRAS by FDA on the basis of a proven history of safe use in food enzyme production (FDA, 2003). These enzymes were already in the marketplace in 1958 when the FDA introduced their GRAS system. The first generation of recDNA strains derived from this strain were two phytase production strains and a xylanase overproducing strain. They were made by randomly integrating expression cassettes for the genes of interest, under the regulation of the host-own glucoamylase promoter.

The newest generation of strains however is based on a "Design & Build" technology. The details of this technology are explained in more detail elsewhere (van Dijck et al, 2003, and references therein). This allows for genes of interest to be over-expressed by integrating them in well-defined and prepared loci in the genome of the host. The method of construction ensures that no heterologous markers such as antibiotic restriction markers or heterologous selection markers are present in the final production organism. Furthermore, due to the fact that the integration is targeted and not random there is no possibility of pleiotrophic hazardous effects, such as the activation of any silent mycotoxin genes mentioned above. Since the genes to be over-expressed are put under the control of the host-own glucoamylase promoter the production process for each new enzyme to be produced can be kept constant. This has major consequences for determining the safety of the resulting enzyme product.

Consumers safety of enzyme preparations is determined by three variables: the producing organism, the raw materials used in the production and the production process itself. The latter one is embedded in current Good Manufacturing Practice (cGMP) and Hazard Analysis of Critical Control Points (HACCP); therefore the safety focus can be directed to raw materials and the producing organism. The use of a strain with a history of safe use and targeted integration according to the concept described above has consequences for the safety studies on the final product. If a known enzymatic activity is over-expressed the safety of a new enzyme preparation is covered by the results of the safety studies performed for other strains from this specific *Aspergillus niger* strain lineage. An overview is published on the available toxicity tests with these strains (van Dijck et al, 2003). For new enzyme products produced with strains from this lineage using the design and build technology no new sub-acute/chronic oral toxicity studies therefore are needed. This has the benefit that no longer test animals are needed to demonstrate the safety of products produced by these strains. Besides this benefit for the test animal there also is the benefit of high expression levels of the enzyme (less costs of raw materials and energy; less production of waste).

The technology also allows to over-express genes of interest from low-producing, less common or less safe sources and thus avoids spending valuable resources on developing production processes for such sources.

8. Concluding Remarks

- Industry responds to pressure of government (taxes), provided there is enough time to develop adequate solutions, as can be deduced from the case of the waste water purification process.

- The driving force for innovation is competition for global market share. Governmental policy should be aimed in alleviating obstacles for innovation, in conjunction with stimulatory measures. In Europe this even-handed approach has not been followed with respect to the use of GMO technology over the past decade.

- At least for products and processes from industrial (also called White) Biotechnology we can claim that there is a strict correlation between innovation and benefits for the environment. All cases in this paper underline this conclusion

- Whether or not innovation by the use of rDNA technology in the food sector will lead to products accepted by the consumers is largely dependant on the final two players in the food chain: the food processors and the retailers. Environmental benefits of White Biotech innovative processes thus are also dependent on their willingness to inform the consumer at large about the benefits of the technology for society.

- Research at DSM is directed towards developing products and processes in response to needs and opportunities in the marketplace. For this they use all relevant techniques that are available to them. It is product specifications and time to market that determine which techniques will be used. In many instances this will be a recombinant DNA approach, but there are definite cases where traditional strain breeding techniques still prevail.

References

(1) Anonymous. A fresh route to an antibiotic precursor. *Chem. Eng.*, October 2001, p. 94.

(2) Burns M, van Dijck PWM. The development of the penicillin production process in Delft, The Netherlands, during World War II under Nazi occupation. *Adv. Appl. Microbiol.* 2002. 51: 185-200.

(3) Burns M, Bennett J, van Dijck PWM. Code name Bacinol. *ASM News.* 2003 69: 25-31.

(4) Elander RP. Industrial production of β-lactam antibiotics. *Appl. Microbiol. Biotechnol.* 2003 61: 385-392.

(5) FDA GRAS notice no. GRN 000089, April 3, 2002

(6) Hersbach GJM, van der Beek CP, van Dijck PWM. The penicillins: properties, biosynthesis, and fermentation. In *Biotechnology of Industrial Antibiotics* (EJ Vandamme, ed) Marcel Dekker, New York 1984. pp. 45-140.

(7) Schuster E, Dunn-Coleman N, Frisvad JC, van Dijck PWM. On the Safety of Aspergillus Niger – a Review. *Appl. Microbiol. Biotechnol.* 2002 59: 426-435.

(8) Sijbesma F. White biotechnology: Gateway to a more sustainable future. *Biovision Congress Lyon* April 10, 2003.

(9) Van Dijck PWM. Chymosin and Phytase. Made by genetic engineering. *J. Biotechnol.* 1999 67: 77-80

(10) van Dijck PWM, GCM Selten, RA Hempenius. On the safety of a new generation of DSM *Aspergillus niger* enzyme production strains *Reg. Toxicol. Pharmacol.* 2003 38: 27-35.

THE AGRICULTURAL SECTOR:
IMPACT ON AGRICULTURAL MARKETS AND COMPETITIVINESS

Sip S. de Vries
Committee on 'Biotechnology, Bioenergy and Agricultural Resources'.
COPA and COGECA.
Rue de la Science 23-25.
1040 Brussels. Belgium

1. COPA and COGECA

COPA represents the farmers of the European Union as they are organised in several farmers' organisations in the member states. COGECA is the representative body for the farmers' co-operatives in the member states. The two bodies have their secretariat in Brussels and organise many Committee that deal with agricultural issues.

One of them is the Committee for 'Biotechnology, bio-energy and agricultural resources' that was established during the mid eighties of the last century. This naming of the Committee clearly indicates that biotechnology is interpreted as potentially instrumental to agriculture and processing industries in agriculture. Biotechnology can hardly be judged as such, *pur sang*, and it should not be judged in such a way.

2. European Commission expert groups

ETAP-SPC
The most recent papers that the European Commission released on biotechnology issues, clearly refers in the same way to the instrumental approach to biotechnology. It can be referred for example to the expert analyses in the Environmental Technology for Sustainable Development Action Plan (ETAP) and especially to the issue group that resorts under it and which is called the Issue Group for "Sustainable Production and Consumption" (SPC).

This ETAP-SPC Issue Group has been established as a consequence of the Lisbon Process that entails a comprehensive strategy and action plan to promote the development and application of biotechnology in Europe, as one of the main drivers of European economic growth and competitiveness.

The paper of the Issue Group on biotechnology that was issued in February 2003 is called "Biotechnology as an environmental technology"(6)[1] starts saying that "Biotechnology cannot be said to be inherently sustainable and environmentally friendly because, like practically any technology, the environmental impact of biotechnology varies from one application to another".

The paper goes on by stating that "The most distinctive feature of biotechnology is the use of biological processes in place of reliance on chemical and physical processes. Thus it would be reasonable to say that, in general, biological processes have less of an environmental impact than physical or chemical ones because biological mechanisms tend to favour efficient usage of energy and other inputs, and thus create less waste".

---

[1] The number(s) between parenthesis indicate the specific literature reference as listed at the end of this paper.

Scientifically seen this statement is of course too general and should at least be proven for the sake of consumer trust. It seems at least very hard to verify it in general. The statement also denies the issue of 'risk and opportunity' which is so intriguingly present for the consumer when talking about applications of biotechnology.

However, the quoted statements of the ETAP-SPC issue group also indicate quite clearly that talking about risk *only* is not that clever. Only risk assessment and developing methodologies to do such, cannot be the single issue when biotechnological applications are at stake.
Apart from 'risks', 'risk assessment' and scientific ways to assess risks, we clearly have to emphasize 'opportunities', 'opportunity assessment' and scientific ways to assess opportunities of biotechnological applications.

3. European Technology Platform

Another initiative of the European Commission on the issue of opportunities of biotechnology started with a Workshop to establish "A European Technology Platform on Plant Genomics and Biotechnology". The Workshop has been a result of the call from the European Council in March 2003 "to strengthen the European Research and Innovation Area…in the enlarged EU by creating European technology platforms… to develop a strategic agenda for leading technologies…".

The objectives of the workshop are:

-       to disclose the potentials of genomics and biotechnology for addressing social, economic and environmental challenges;

-       to discover challenges to encourage the use of genomics and biotechnology;

-       to identify interested partners and identify the time table for progress in this field.

4. Risks and Opportunities

Opportunities of biotechnology to encounter existing risks of existing processes are referred to in both initiatives of the European Commission that are quoted above. It should be stressed in this context that the European Parliament in its recent decisions of 2 July on genetically modified food and feed [COM (2001) 425] and on traceability and on market regulation [COM (2001)182] only names risk assessment and does not talk about opportunity assessment of biotechnological research and applications.

The same is (*partially*) true for the European Commission. In a recent press release of the European Commissioners Wallström and Byrne, as a reaction to the voting in the European Parliament, also referred to risk assessment and less than before to opportunities. The Commissioners state that 'Clear rules are set out in the EU for the assessment and authorisation of GMOs and GM food, but responsibilities are shared between Member States and the Community. The scientific risk assessment will be carried out by the European Food Safety Authority…The regulation also establishes the Joint Research Centre (JRC) of the Commission as new Community reference Laboratory, which will have the main task of validating detection methods".

5. Co-existence of Cultivation Methods as an Issue

This risk assessment on a life cycle basis is accepted by COPA and COGECA and the authorities named are able to perform such a task. The *condition of co-existence* that COPA and COGECA have pushed so severely, will not be easy to control, but the Joint Research Centre is equipped with very good scientists at work everywhere in Europe.

Co-existence of different agricultural cultivation methods presents more than one aspect. Another issue is of course, whether new technologies and especially patented bio-technology will be equally accessible for farmers. Instead of *detectability* of biotechnological inventions, this co-existence theme deals with *accessability* to biotechnological inventions.

This latter theme should not be forgotten, although at the moment, it is not central in the discussions. Patent rights opposite to breeders rights remain an issue to be addressed, because of the accessibility to new technology.

However, risk assessment which guarantees co-existence is one side of the coin, although a very important one from the view of primary agriculture. It is a *necessary first step* to introduce biotechnology in the food chains and to get the consumer acquainted to food that in one way or the other has obtained traces of modern biotechnology and to the safety-ness and the reasonability of these uses and applications.

Co-existence is a necessary condition now, but may develop in two directions in the future, namely:

a.   co-existence develops such that different methods of agricultural cultivation will get more explicit to the consumer;

or

a.   it will become a vague concept in the future because a more widespread mixture of agricultural cultivation methods will develop in which biotechnology will have its proper place.

6. Elements that Push, Pull or Block Developments

Which direction will be taken highly depends on elements that will pull and push, and moreover on elements that will block further developments.

The *pull elements* refer to the market possibilities, which means the market prices and costs of the products that entail biotechnology versus the market prices and costs of products that do not or only to a certain threshold entail biotechnological traces.

Moreover, *pull elements* refer, at the end, to the acceptance of the consumers and consumer groups. At this very moment, we cannot say a lot with certainly about this aspect for the extended European Union of 25 member states.

The proof of the pudding, will be the in the eating by certain consumer groups. The question of course is whether the 'pudding' should be subject to explicit marketing or not.

Figure 1. Push, pull and blocking elements to biotechnology development

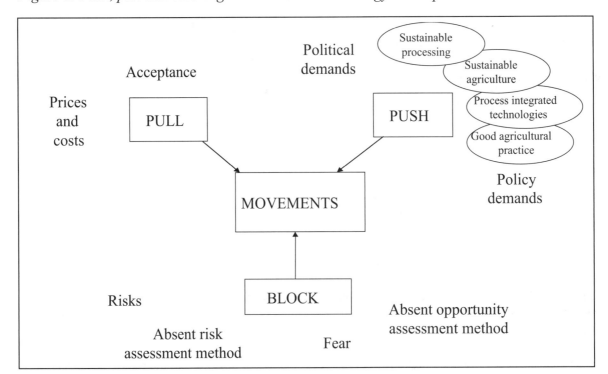

The *push elements* clearly refer to political and policy demands, whether from public authorities or from inside the agricultural sectors or sector-chains themselves.

Push elements will be for example dependent on:

-     the definition of 'good agricultural practice',

-     the definition of 'process integrated technologies',

-     the definition of 'sustainable agriculture' and

-     the definition of 'sustainable processing'.

Clearly these concepts are constantly re-considered, re-weighted and re-established, also because of developments in biotechnological science itself.

Apart from pull and push elements we have to recognize the *blocking elements*. These are factors that *block* proper development and application of biotechnological inventions.

'Fear' and 'absence of proper information' are the main blocking elements to new technologies.

What we should strive for is the further development of scientific (laboratory) methods and in-field tests that enable *the simulation* of longer term practice of biotechnological applications and of their effects.

Moreover, the existence of such scientific simulation tests should be stressed in all discussions with interested groups.

What could be a major blocking element to the development and use of biotechnology is '*the absence of opportunity assessment*'.

The European Commission is working very hard on this issue at the moment, as I indicated before. It is absolutely necessary to develop methods that enable *the assessment of opportunities* and the possible contributions of these new technologies to problems that are encountered at the moment. The general public should be informed widely about this *assessment of opportunities.*

7. Opportunities and Assessment

Many people and organisations work quite hard to communicate the benefits of specific biotechnological innovations. Ten years ago the opportunities of biotechnology were presented to the public in a very general way, as Figure 2 on 'possible biotechnological applications' shows. Biotechnology at that time was supposed to be applied *everywhere* and *very soon*. It was presented as 'un-escapable', 'necessary', 'globally applicable', 'uniformly problem solving'. This view on Biotechnology of 10-15 years ago has *as such* increased the fear for applications of biotechnology, because it resulted in quotations of 'Brave New World' that were presented as reaction.

Figure 2 Opportunities detected in the late eighties and early nineties

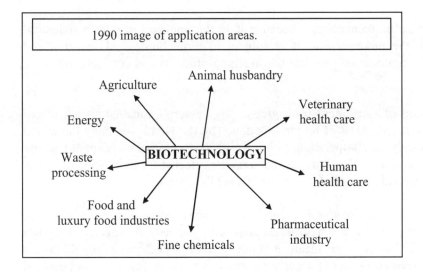

Nowadays biotechnology is presented differently, as the activities of the European Commission indicate. The development and application of specific biotechnological inventions in specific areas of interest is now central.

As figure 3 shows, currently a clear view *where* biotechnology and also *what* biotechnology has to be applied is developing in order to counter currently experienced hot issues. Therefore, if one decennium ago no fixed idea about where or how to apply biotechnology existed, nowadays, there is a clear demand for specific application in those areas that can be detected as highly energy consuming and areas that are fitted for substitution products or processes.

Figure 3: Opportunities of biotechnology detected nowadays

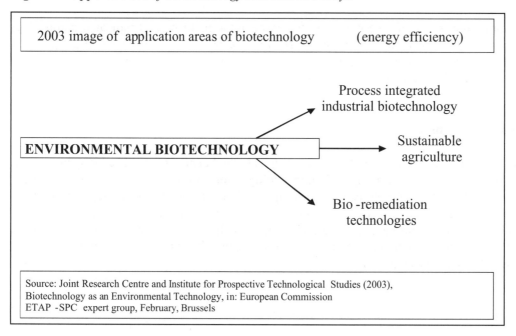

This is also called sometimes 'white biotechnology', because it is the application within industrial processes that are separated from the environment. It is subject to biotechnological research and application because current production processes generate too much co-products and consume too much energy.

A striking example is the production of lactic acid from glucose by bacterial conversion, where co-products that are not wished are produced. A lot of research is done (by the Dutch company Purac and the American company Dow Chemicals, in co-operation with Cargill) to develop new, adapted bacteria that enable a much more specific conversion. New, adapted bacteria are selected for more specific, purer and quicker conversion. Energy saving and absence of waste streams are the criteria.

Another example is the conversion of cellulose to C6 sugars and the conversion of hemi-cellulose to mainly C5 sugars. When we are able to do this hydrolysis step with the help of specific enzymes (reorganised amino acids) or specified bacteria, this means that new feedstock becomes available for the production of bio-chemicals. These feedstocks were formerly regarded as waste or low value agricultural co-products, like sugar beet bagasse or rice straw and hulls.

It is not surprising that research all over the world looks for enzymes and bacteria, that enable proper, pure and energy saving conversion of cellulose feedstock to ethanol. Recent developments are genetically engineered bacteria [(Escherichia) Coli KO11 and Erwinia chrysanthemi EC16] for conversion of the carbohydrates galactose and arabinose.

Such biotechnological inventions can normally only be applied when other technologies are applied as well (like 'wet air oxidation' on a cellulose feedstock that contains a lot of lignin). This once more demonstrates that biotechnology is not a technology to be developed separately, but in conjunction with other technologies. This is what it normally meant with *'Integrated Agro- Food Systems'*.

The essential aspect of such developments is to try to reach higher yields by more specific working organisms. This means, organisms that reduce the conversion of sugars to carbondioxide and water, while increasing the production of products that are less oxidised than carbondioxide.

Another essential element of such developments is that process conditions can remain nearly environmental (atmospheric pressure and normal temperatures). The opportunities for biotechnology (mostly enzymatic because they act as catalyser in processes) or new technologies in industrial processes, can be assessed by discovering sub-processes that do currently only take place under severe process conditions (mostly high temperatures and high pressure conditions).

9. Sustainable Agriculture

If we like co-existence of different agricultural production methods, space should be given to the development of 'precision agriculture'. From an environmental, energetic and an economic point of view, the reduction in the use of nutrients and crop protection means by agricultural crops, is a main issue in agriculture. If adapted and more specific breeds of crops can partly achieve this, it will be welcomed given proper risk assessment and opportunity assessment.

An opportunity assessment provides us with the *positive* effects of introducing adapted crops in terms of:

- reduction of fossil energy use, nutrient use and pesticide use;

- adaptation to different climatic conditions, and, therefore, possible changes in cultivation schemes;

- resistance to diseases;

- efficiency in water use or consumption during the growing season;

- contribution to long term sustainability;

- economic benefits.

There are many other developments that can be named, such as the production of substitutes for polypropylene in the plant system itself. This means that the breed is conditioned in such a way that it activates a conversion of sugars in certain stage of growing. This research has been already undertaken for a long time at the University of York.

10. Bio-Remediation

Bio-remediation is a method of dealing with waste streams of different processes. It is quite clear that biotechnology can be very helpful in this field. Researchers and public authorities in the European Union are currently dedicating a lot of attention to the issue of how to deal with minimisation and valorisation of waste streams.

11. Developing Methods to Assess Risks and to Assess Opportunities

The necessity to develop scientific methods that are not only able to assess risks, but also enable us to describe and to assess the benefits and opportunities in a environmental and an economic way should be stressed.

There is also the necessity to develop such methods, that are able to identify 'process losses' of current activities and thus opportunities to reduce them by adapted technologies. It is necessary because we want to clarify to consumers what the environmental end economic interest will be. At the end this will convince the consumer about the specific value of new technologies on specific locations.

Figure 4: The issues in an opportunity assessment

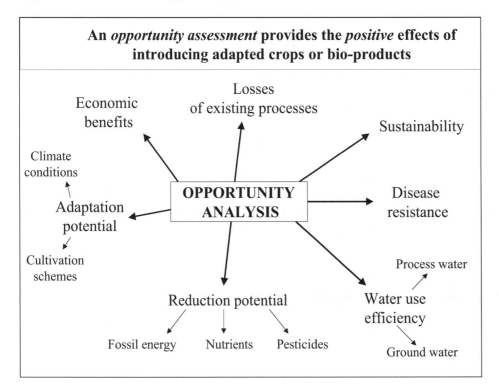

An opportunity assessment, as indicated in figure 4 above, provides an analysis of the positive effects of introducing adapted crops in terms of:

- reducing losses in current processes

- reduction of fossil energy use, nutrient use and pesticide use

- adaptation to different climatic conditions

- resistance to diseases

- efficiency in water use or consumption during the growing season

- contribution to long term sustainability

- economic benefits

It is such an opportunity assessment that is complementary to a risk-assessment.

References

(1) CEC (2003), Commission Cummunication COM (2003) 226 on 'Investing in research: an action plan for Europe', Brussels (http://europa.eu.int/comm/research/era/3pct/pdf/action-plan.pdf)

(2) COPA and COGECA (2000), Position of COPA and COGECA on the use of Gene Technology in Agriculture, 21 January, Brussels

(3) COPA and COGECA (2000), Position of COPA\and COGECA on the Communication on the Precautionary Principle [COM (2000) 1], 9 June, Brussels

(4) COPA and COGECA (2002), Remarks on the draft regulation of the Council and the European Parliament on Genetically Modified Food and feed [COM (2001) 425] and Draft Regulation of the Council and the European Parliament concerning the Traceability and Labelling of Genetically Modified Organisms and the Traceability of Food and Feed Products produced from genetically Modified Organisms and amending Directive 2001/18/EC [COM (2001) 182], 8 March, Brussels

(5) COPA and COGECA (2003), Comments on the Communication from Mr. Fischler to the Commission on the Co-existence of genetically modified, conventional and organic crops, 19. May, Brussels

(6) JRC and IPTS (2003), Biotechnology as an Environmental Technology [Working document of Environmental Technology for Sustainable Development Action Plan; Sustainable Production and Consumption Issue Group], February, Brussels

(7) Vries, S.S. de (1987), Biotechnology: Chances and Threats for the Dutch Agricultural Sector; Dutch Agricultural Board, January, The Hague

(8) Vries, S.S. de (1988), Biotechnology: the farmers' interest, in: World Farmers' Times

(9) Vries, S.S. de (1988), Biotechnology and Agricultural Cooperatives, in: Dutch Monthly Cooperation

(10) Vries, S.S. de (1989), Who is the small farmer, in: Trends in Biotechnology, Vol. 7, no. 1

(11) Vries, S.S. de (1991), Biotechnology for Agriculture; The interest of farmers in the discussion and the impacts, CEA Conference, Varzim

(12) Vries, S.S. de (1992), Biotechnology and Agriculture, CEA Conference, Stockholm

(13) Vries, S.S. de (1999), Using thermodynamic principles to compare agricultural chains, Österreichische Gesellschaft für Biotechnologie, Graz

(14) Vries, S.S. de (1999), Thermodynamic and Economic Principles and the Assessment of Bioenergy, The Hague

A FARMER'S PERSPECTIVE ON GMOs RISK ANALYSIS

Jerry Ploehn
National Corn Growers Association. Biotechnology Working Group
Minnesota Corn & Research Promotion Counsel
Ploehn Farms.
85703 600th Avenue
Alpha, MN 56111. USA

Abstract

The U.S. farmer's perspective on biotechnology is well summarized up by two simple statistics:

1) Since the introduction of Roundup Ready soybeans in 1996, plantings have grown to 75% of the U.S. soybean crop.
2) Biotechnology enhanced corn varieties are now planted on 34% of U.S. corn acres.

This rather rapid rate of adoption of new technology is driven primarily by a desire to produce more efficiently, by increasing yield per acre or reducing pest control costs. American farmers generally cite these two factors, along with a reduced introduction of pesticides into the environment, as benefits of GMO crops.

Increased yield, however, is considered to be a benefit for some, and a cost for others. Those who consider it increased yield a cost generally accept that crop prices are reduced as supplies grow larger. Those who consider it a benefit are those who either add value by further processing the crop, such as livestock or ethanol producers. These producers realize greater profits by capturing greater processing dividends per acre due to higher yields. Another group that benefits includes producers in the furthest west and north reaches of the Corn Belt. Herbicide tolerant crops have essentially made it possible for many producers in the region to grow soybeans and corn on land previously restricted to wheat, barley, hay, and pasture. This has allowed them to diversify their operations and raise higher valued crops.

Reduced pesticide costs are realized when treatments are rendered unnecessary, or when more economical pesticides can be used. Many farmers consider the benefit to the environment from reduced crop protection product use. This cost is somewhat offset by a fee to use the GMO technology.

Consumer acceptance is another risk consideration for producers. U.S. consumers generally consider other characteristics such as food safety and bacteria or other contamination as much more important than GMO content. Most consumers express confidence in our regulatory agencies ability to ensure food safety.

Crops genetically modified to adapt to challenging growing conditions, such as excess or inadequate water, high or low pH or salinity will further change the picture for producers. As indicated earlier, these adaptations will be benefits for some, and reduce profits for others.

Lastly, the producer bears a responsibility to manage these new plant products properly. This will become even more pronounced with the development in the near future of crops that contain pharmaceutical components or nutritional enhancements. Segregation of plant material, both growing and in storage, as well as securing proper markets for these crops, will be crucial to long term acceptance.

1. A Farmer's Perspective

Thank you for the opportunity to share my farming operation with you. I will try to describe and explain my farming methods to you. Each farmer in the United States makes decisions based on what the farmer feels is right for his or her farming operation, and this differs from farm to farm.

I will address the benefits and concerns of using biotechnology. Please form your own thoughts about this subject. I am not telling others how to farm, only explaining how I farm.

Our farm is located in the southwest corner of Minnesota (Fig. 1). We raise corn and soybeans on about 400 hectares. This is an average size for a farm in the Midwestern United States.

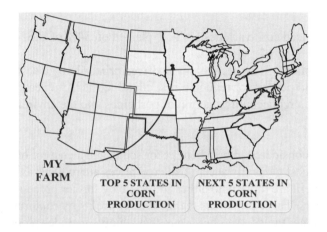

Fig. 1

The top five corn-producing states in the United States are Iowa, Illinois, Indiana, Minnesota, and Nebraska (Fig. 1). These states produce two-thirds of the United States corn crop each year. The other five states noted are Kansas, Missouri, Ohio, South Dakota, and Wisconsin, and together these ten states produce 85% of the United States corn crop.

I have been farming for 41 years and when I started farming, I used primarily organic farming methods. However, I ran into some problems. I encountered plant diseases, such as rust in my oats, and crop insects such as spider mites in soybeans and corn borers in corn. Spraying with chemicals was the only way to control these diseases and pests.

Sprays were an effective tool to use in my farming operation; however, sprays also have disadvantages, which I will mention later. Now we have biotechnological methods of controlling these problems, which is one reason why my farming practices have changed over the years. All of these methods are important to farmers.

Fig. 2

Figure 2 shows aerial views of my farms. The left picture is my home place, and the picture on the right is a view of my grain bins (silos). In the Midwestern United States, many farmers live in the country with fields surrounding their houses. You will notice in the picture of our grain bins that I have a number of grain bins. This enables me to keep conventional and Biotech crops separate.

Fig. 3

My wife, Lorie, gives presentations to first-grade students and their families to explain the numerous products of corn and soybeans. We, as farmers, realize the importance of educating our consumers and we are also aware of misconceptions. By inviting people to our farm, we share with them the uses of corn and soybeans, as well as answer any questions the students or their families have. They are amazed at how many products contain corn or soybeans and at the ways corn and soybeans can be used. We discuss conventional farming, biotechnology, and organic farming.

On our farm we plant some biotech crops and some conventional crops. We feel each kind has certain advantages.

The field-level benefits most cited for biotech crops are:

1. *Greater efficiency of production.*
Efficiency has increased because yields per acre (or hectare) are greater.

2. *Reduced use of more toxic crop protection products.*

Crop protection products are used less (in the case of insecticides on Bt corn) or are less toxic (Roundup).

3. *Reduced tillage required, resulting in reduced soil erosion and lower fuel consumption.*
Tillage reduction lowers erosion, labor costs, and fuel consumption as well as the resulting air pollution.

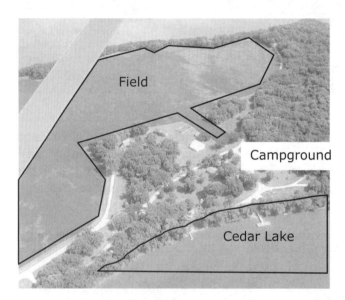

Fig. 4

Figure 4 is an aerial picture of one of our fields. This field has a lake on one side and a public campground next to the lake on the other side. In this particular field, I prefer to use biotech crops because I do not have to spray biotech crops. I do not like to spray this field because of all the visitors and their pets staying in the campgrounds. Of course we obey the rules and regulations about spraying, but animals can run in and out of the fields at will. Then the families play with their pets or the animals run into the water. Sprays are safe to use when used properly, but in this case a better alternative is to use biotech crops, so I lessen my need to spray. As Minnesota is known as the "Land of 10,000 Lakes", we are very careful to avoid spray drift and run-off.

Earlier I described some field level benefits of biotech crops. Now I will explain field level costs. These are:

1. *Increased input costs.*
Biotech seeds are priced according to what the market will bear. The reduction in use of other pesticides is not always great enough to offset the added seed cost.

2. *Increased yields can lower crop prices.*
      i. Shifts in competitive advantage.
- If you had a competitive advantage in corn production because you farm in an area with traditionally low insect pressures, Bt corn use in a different area can negate your previous advantage.
- Another good example is the use of Roundup Ready Soybeans, which have enabled soybean production to move into non-traditional soybean production areas.

ii. Consumer acceptability.

Consumer issues are not a large factor in the United States, where people generally have confidence in our regulatory agencies to monitor food safety. Consumers are still much more concerned about food safety issues resulting from bacterial contamination, food freshness, and handling than are about biotech crops.

3. *Producer responsibility.*

- The producer has long been aware of his responsibility to use varying types of crop production products to prevent pests from developing resistance. This new technology represents another level of responsibility.

- The producer is also responsible to segregate specialty trait grains to assure that they are used in appropriate markets. This will become even more important with the advent of new traits such as pharmaceutical or nutritional enhancements. As a producer, I believe that the marketplace will attach a premium to these new products, creating an incentive for me as a farmer to adequately segregate different types of grain.

On the following four pages, I have included maps showing corn production and soybean production in the United States. You will see that between 1994 and 2002, crop production increased and that the production areas have shifted slightly. Some of this is due to the use of GMO crops. GMO seeds have enabled farmers to raise crops in non-traditional production areas

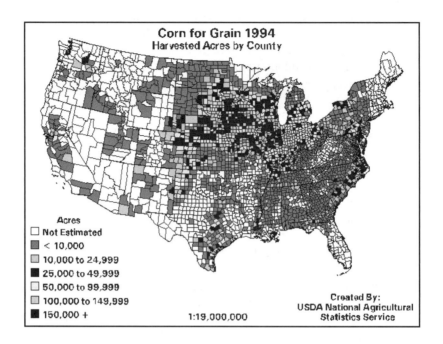

Top Map: Figure 6   Bottom Map: Figure 7

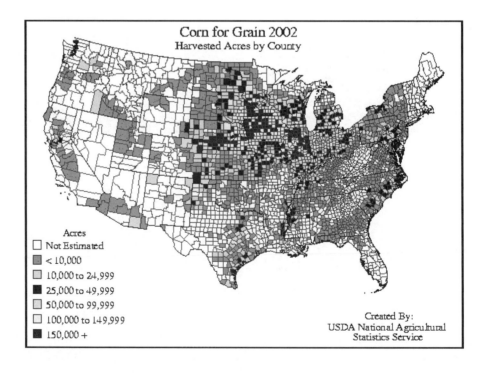

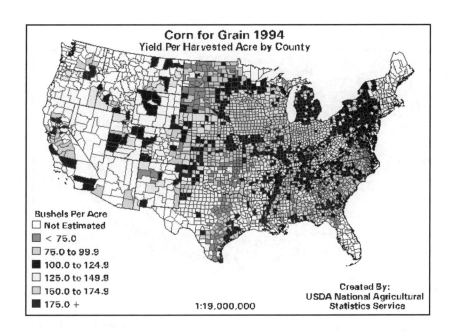

Top Map: Figure 7   Bottom Map: Figure 8

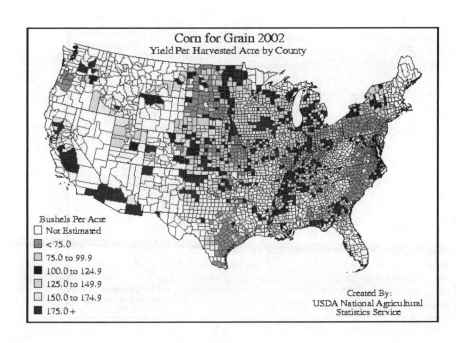

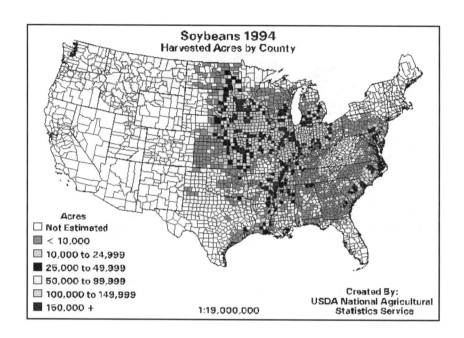

Top Map: Figure 9   Bottom Map: Figure 10

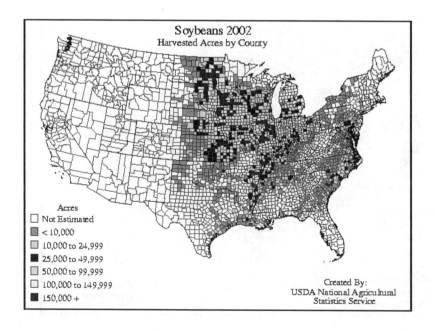

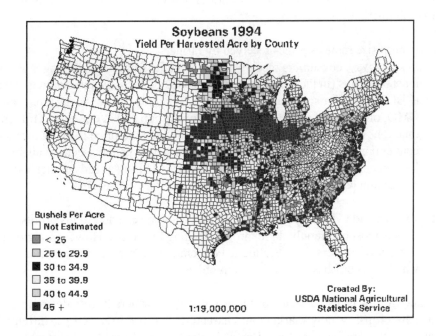

Top Map: Figure 11   Bottom Map: Figure 12

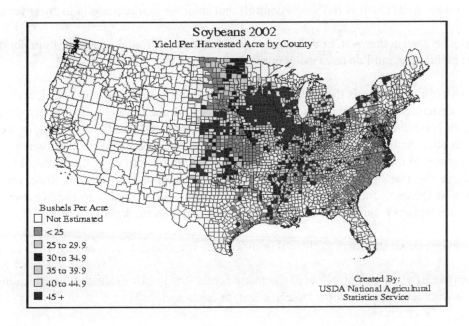

The preceding maps have shown production of corn and soybeans in the United States. There are two sides to increased production:

1. Increased crop size reduces prices received by farmers

    In some way, consumers receive hidden benefits from biotechnology. For instance, most soybeans (in the world) are priced based on the Chicago Board of Trade's (CBOT) price for soybeans. CBOT's price includes all soybeans (conventional, GMO, or Roundup Ready Soybeans) in one group and at one price. If there are more total tons of soybeans produced in the world, then the price is lower for both conventional and GMO soybeans. Also, the newer genetic seeds available tend to have better yields than the seeds of yesterday. This also leads to increased production worldwide, which lowers the price farmers receive.

2. If the farmer can add value through further processing, he can add to his profits.

    This is most commonly done through feeding livestock. However, a growing number of farmers are investing in cooperatives to process their corn into ethanol, which is a corn-based fuel for automobiles.

Production of "new" crops is also a factor of biotechnology. Farmers have a desire to diversify and they are looking for new crops, or new marketable characteristics in current crops, through biotechnology. Some new characteristics are increased nutritional content, special health characteristics, pharmaceutical proteins, and the list goes on. As a farmer, I am hopeful that these new products will help meet consumer needs around the world, as well as the farmer's bottom line. Through careful consideration at all levels, from research and testing to commercialization, I believe we can meet these goals.

Today, one-third of all corn and three-fourths of all soybeans and cotton in the United States are GMO crops. Producers (farmers) are generally satisfied with genetically enhanced crops. Producers anticipate that new output traits (such as rootworm control) will enter the marketplace in the near future.

However, there is another area in the subject of GMOs that is of vital concern. I am not qualified to give opinions in this area, but I do have many questions.

The area I am referring to is the reality of GMO agriculture; the economic, social, and political realities of GMO agriculture. Many farmers throughout the world are finding it extremely difficult to make a living from agriculture. This holds true for large American farms and family farms, as well as very small farmers in other parts of the world. What does GMO agriculture mean for their way of life? Is it good for a company to make large wind-fall profits and only share this reward with top-level management? What about the rest of the employees? Should they receive a portion of these rewards? Did the customers of the companies pay too much? Should consumers be able to receive more value for their money? Are farmers receiving a fair value for the crops and livestock they produce?

I believe everyone should share fairly in this endeavor. The question is, what is FAIR?

I have only briefly touched on a few of the many issues that I must evaluate when deciding what to plant. Do I farm with conventional crops, biotech crops, or organic crops?

Thank you so much for allowing me to share my farming operation with you and for taking time to understand it. I wish you well in the future.

THE CONSUMER: THE RIGHT TO INFORMATION AND SPECIFIC REQUIREMENTS ON STRATEGIES

Beate Kettlitz
BEUC. The European Consumers' Organisation.
Avenue de Tervueren 36/4
1040 Brussels, Belgium

This paper was presented orally, but was unfortunately not available at the time of publication.

CHALLENGES FOR THE MEDIA: DISSEMINATING INFORMATION BY AVOIDING HYSTERIA

Katherine Williams
Editor, AgraFood Biotech – Agra Europe
80 Calverley Road, Tunbridge Wells
Kent, TN1 2UN, UK

Abstract

The media faces numerous challenges in attempting to inform the public about real hazards and dangers in this world without causing mass hysteria.
These challenges include the audience that is targeted, selection of appropriate language, and finding a balance between points of view. Other challenges that face the media include identifying the stories that genuinely require attention and recognising those based on hype or false claims.

Reporters are under a number of pressures from the media itself and this can lead to conflicts between the hysteria driving viewpoint and the more objective position. We must all be aware that bad news sells, conflict and controversy drive the news, and ratings or sales are important. Reporters also work to deadlines, and in an increasingly Internet driven age there is more pressure than ever to meet time restrictions.

The media is also faced with manipulation from a variety of sources, pressure groups, big business, political parties and so on. This can affect the quality of coverage surrounding an issue.
It must also be kept in mind that since the media are reporting for the public, the public have a duty to carefully analyse the information they receive, hysteria is not usually the result of a journalist's comment but of the public reaction to it. People do not base their decisions on the media alone, they do not believe everything they read and a great may other social factors come into play regarding hysteria.

1. Introduction

Hello, well the first thing I'd like to say is that despite the headlines promising nuclear terrorism, engineered viruses, rogue machines and grey goo we may not be doomed - so please don't panic.

I have been asked to talk about Challenges for the Media: Disseminating Information by Avoiding Hysteria.

I feel that in many ways the challenges facing the media are similar to those faced by researchers:

- Both want the public to know about new discoveries - the scientists partly because publicity helps with research funding and journalists because they want to sell copy!
- Both try to be objective
- Both face the challenge of communicating new results, which means they may have to educate the public to some extent.

The main difference in our challenges as I see it is that scientists tend to write for other scientists. Whereas the media has a different audience - the public. This brings us to a difference in types of media the specialist press verses the popular press.

The general public is confused and fed up with contradictory news reports over the pros and cons of new and old technologies.

They want a straightforward "Yes this is good" or "No this is bad." This presents a challenge for the writer since all to often the answer (as all things in life) is that there are some benefits and some risks.

Journalists often source their stories by rewriting press releases, adding some quotes from people involved and people who have an opposing view for balance. In the case of a newspaper their sub editor may then cut out 500 or so words to fit the story onto the page. This may just happen to be the "key" 500 words that avoided a panic or balanced a quote or explained the limitations of the study.

2. Role as Educators

The specialist press has a reduced educational role, we can usually assume that since someone is reading a specialist publication they know a little about the topic.

However, in the case of the popular press the reporter will have to deliver an oversimplified message, often the public does not understand the details behind a scientific method or a statistical analysis and some form of explanation will have to be offered. For example dozens of health and nutrition stories make it into the news media on a daily basis. Unfortunately most have no valuable take home message for readers. A good research study raises more questions than it answers. This is why so many scientists add "but we need to do more research."

The story should tell the reader how many people were studied. If only 20 people were in a group, the findings are less important than if 5,000 were analysed. Also, intervention studies (in which someone is told to change their diet or take a medication) tend to be much more reliable than an observational study in which people report on their habits with greater or lesser degrees of accuracy - can you remember what you ate five years ago?
Also a study carried out using healthy volunteers may not have the same importance for a patient with a disease. Analysis of results in men may not always apply to women. Generalisation of the results can lead to a hysterical response.
Of course, this method of hysteria avoidance relies on journalists understanding the analysis themselves.

3. Language

Selecting the appropriate language can be challenging. Language can be enlightening but it can also mislead through suggestion or use of emotional overtones. We only need to look at the words associated with the GM debate for evidence of this "Frankenfoods," "genetically altered" and "mutant" sound far scarier than "biotech" or "genetically modified." Even though the word mutant is a scientific term it has been popularised as a frightening thing.
In addition words such as "may," "might," "could," "possibly," "perhaps," and "potentially", are all capable of causing panic if strategically or carelessly placed.
However, language can also have unexpected results, during the recent SARS events risk communications experts say the media's anti-panic response made the public *MORE* likely to panic. They

discovered that statements such as "everything is under control" while appearing reassuring actually frightened the public. This was thought to be because the public concluded that if closed hospitals and multiple face masks were a sign of a situation that was "under control" then someone, somewhere was lying and no one was acknowledging that fact. This is a challenge that can be very difficult to face. In this case people needed to hear what was being done and why, rather than differing opinions on the outbreak or wild theories about where the disease had come from.

David Letterman joked on his Late Show, "We've got SARS, mad cow disease, an orange alert - the news is so bad the New York Times doesn't have to make it up," referring to a the recent scandal at the newspaper in which a reporter was found to have faked sources and datelines.

It may have been a joke but this comment reflects the fact that bad news sells.
Reporters are pressured to find controversy. Good news stories are read and forgotten. Conflict and controversy drive the news, so sometimes the media would rather generate a scare than present the slightly less interesting balanced view.

Perhaps another important message from the popular press is that journalists can also be victims of spin, they must toe the line of their newspaper or editor and keep advertisers or other funders happy.

Some newspapers or TV channels want headlines to reflect their social and political agendas.

4. Time

Reporters work to deadlines, and in an increasingly Internet driven age there is more pressure than ever to turn a story around quickly. This reduces the opportunities for journalists to provide an objective view. Chasing quotes and fully understanding what is driving an issue can be a challenge. Accepting the quick, hysteria driving version is simple and effective - you meet your deadline - everyone is happy.

This is another case where the specialist press has an advantage - they often realise that a story does not *HAVE* to be presented today. They have more time to research the subject or more knowledge of what has taken place in the run up to a story. They are more willing to tie up the loose ends while the popular press often abandon a story mid-way through.

It should also be kept in mind that those people interested in generating hysteria are very good at manipulating the media, these groups can be all too aware of deadlines and can present a journalist with a ready made story, bite size quotes and interesting photographs.

The reporter may then be too busy, or too lazy to go and track down someone to present the opposing view. Also the person to supply an opposing view may want time to prepare their argument, or to gather evidence to support their view.

5. Who's Who?

Journalists are taught to cover both sides of a given issue, however, when conflicting opinions are presented they are often blandly stated, "He said," "She said," without any regard for who is making the claim, or the scientific consensus behind the claim.

Lack of understanding on the part of a reporter can seriously damage the quality of reporting by the news media, the reporters may fail to investigate or understand whether a so-called expert is an expert in the correct field. A scientist acting as a falsely proclaimed expert, criticising research about which they know little can also contribute greatly to hysteria.

The media must face up to the challenge of recognising what is driving the news crossing their desks:

- Social activists may present hysterical views to achieve social and political change.
- Businesses may spread scares to beat competitors or boost their product sales.
- Politicians may add to scares to curry political favour or generate publicity; and
- Researchers want to present their theory to the widest possible audience.

On the topic of researchers I might add that sometimes they can also be major drivers of hysteria. Some researchers are so keen to share their work that they produce press releases before the data has been thoroughly checked. Journalists then use this information or abstracts from scientific meetings to build their story. Coverage of non-peer-reviewed work can lead to scares and the spread of inaccurate information.

6. Conclusion

A journalist needs to present a story that is accessible to the public (and to the editor) and it needs to be written up quickly and it has to compete on pages full of dramatic events.

In the face of these challenges it is very difficult to say progress is being made - slowly - that there are no good or bad guys, and that the results have no practical significance as yet.

We should also keep in mind that the media play a vital role in alerting us to real hazards and dangers in this world, and that people do not base their decisions on the media alone, they do not believe everything they read and a great may other social factors come into play regarding hysteria.

SESSION 6:
GENERAL DISCUSSION

GENERAL DISCUSSION

B. Le Gallic highlighted certain definition issues. He expressed surprise about the range of terms and concepts used (e.g. GMO, LMO, GMPs, Conventionally Modified Organism, GMO-free, etc.) and pointed out that this may increase difficulties in achieving a clear and transparent discussion on GMOs.
In particular he noted the difference between US and EU approaches, namely: product (US) vs. process (EU) as well as the related distinction between LMOs and GMOs, the first of which seems to exclude processed product from the analysis (e.g. milk, eggs or meat from animal fed on GMOs; cheese using GM enzymes). He concluded that while products may be similar (following the principle of "substantial equivalence"), it may nevertheless be important for the consumer to have access to information on the process used (e.g. see working conditions issues in the shoes sector; origin of energy; the US "dolphin-safe" label for tuna fishing in the fishery sector).

R. Holzinger stated that the USA position on Risk Management is very clear and sound science based exactly because it focus on the product and not on the process. Moreover, the term Genetically Engineered (GE), used in USA, is very appropriate because every plant is genetically modified. He further pointed out that "biotechnologically derived" would be probably a good term for consumers as it is not scary.

S. Hisano commented that critical differences between the US concept of risk assessment procedures ("substantial equivalence") and that of the EU have been revealed on several occasions as well as in this workshop. However, he continued, both scientists and regulators still continue to assure of their "scientific soundness". This clearly shows that even scientific knowledge is based on context-specific understanding by a certain collectivity. The relationship to be intermediated by deliberate communication/discussion is not just between scientists/experts and lay public, but also between scientists/experts themselves. What is meant here is that science is never monolithic. It is not true that every scientific expertise can come to *THE* same conclusion on issues such as those facing a lot of uncertainties, even if its activity is to be carried out on the basis of "sound science" approach. Even in natural science, there could be a wide range of opinions and perspectives toward the concept of risk. In this sense, we might have to refer to science plurally (i.e. sciences) all the time.

M. Inaba arose the issue of *"Risk perception"*. He noted that there are various views on which principles or criteria can identify more than others "Risk Perception". He noted that the area of concern differs among people and asked how we can correctly understand the different views. Risk perception is also a key factor in determining "Public acceptance". However, we still don't predict or even understand much about public attitudes. More research will be needed in this field.

S. Hisano added that the concept of risk communication is drawn on the grounds that public concerns are considered to be based on misunderstanding or lack of scientific knowledge. He noted that the public is mostly equated with consumers in this kind of discourses and he expressed his doubts whether these two categories are a substitute for each other. He further stressed that what is required to be shown to the public is a lack of sufficient and non-ambiguous data/information and the problematic evaluation of long-term and synergetic effects. Once the public perceive that the scientific community admits such uncertainties, which nevertheless reflective scientists persistently struggle to solve, public concerns could become compatible with public understandings (not necessarily with public acceptance, though). Scientists and administrators are required to communicate with the public mutually and deliberately (and ideally to institutionalise public participation in the process somehow), about the process of scientific evaluation including its difficulties and uncertainties (as did the workshop to some extent), not just about the results of allegedly "sound science"-based assessment. In this regard, a report of the PABE (Public

Perceptions of Agricultural Biotechnology in Europe) research project (funded by the Commission of European Communities) gives us a lot of implications. It reveals that: "Although ordinary citizens are largely ignorant of the scientific technicalities of genetic manipulation, and of developments in research, regulation and commercialisation related to GMOs, this lack of knowledge does not explain their response to agricultural biotechnologies". It also states that their concerns expressed in the focus groups were mostly based on experience-based knowledge about the behaviour of insects, plants and animals, about human fallibility, and about the past behaviour of institutions responsible for the development and regulation of technological innovations and risks. These are only a tiny part of key findings of the research project. Such a deep gap between the kind of knowledge mobilised by the lay public to evaluate GMOs and the kind of knowledge assumed to be relevant by scientists, administrators and promoters of GMOs is an important input from sociology.

B. Rudloff noted that from the presentations and discussions during this meeting it seems that that public perception on food and food safety is very different from the reality. She stressed that we all (scientists, policy makers, and media) have a role allowing the consumer to be aware that nothing is 100% risk free and that we all need to be open about uncertainties. Only through continuous communication among all stakeholders it is possible to come to a well-informed debate. In this context labelling can be a very important tool for communication purposes. However, labelling is not an alternative to public education, people do not know what is in their food or how agriculture works.

R. Holzinger highlighted that, though the new EU labelling rules should have as objectives to inform the consumer allowing free choice and to build his/her confidence, these goals won't be reached because these new rules are inconsequent (therefore will lead to loss of confidence) and one-sided (therefore do not allow a real choice and are misleading). For example the mentioned rules require a label for products containing or consisting of GMSs (wie bisher) and products produced from GMSs (ingredients, flavourings, additives) irrespectively of the detectability of GM DNA or Protein. On the other hand, no label is required GM processing aids (Chymosin) and products from animals fed on GMOs (milk, eggs, and meat). He stated that the lay public still wants to read black/white opinions/recommendations and the EU labels in his opinion do not help in this respect. He closed his intervention on this point asking why do we just label GMO as a breeding technique, while the other breeding techniques don't have to be labelled (e.g. irradiation and vegetative cloning).

A.D. Hartkamp noted that one of the speakers stated that there are no GMOs products on the market in the EU and that there is no demand for them. She pointed out that instead some supermarkets in the Netherlands are *voluntarily* labelling foods derived from GMOs (e.g., corn oil) and since they started to do so, sales of these products have not dropped.

Anonimous participant asked how can the European scepticism on GMOs can be overcome.

K.H. Madsen replied that certainly information and education campaigns (at the European, national and local level) are necessary. Moreover if GM products will prove to be not only 2% cheaper but approximately 10% cheaper than conventional products acceptance should grow.

M. Inaba commented that in his opinion labelling and acceptance are issues which influence each other substantially. He noted that that this will be the major issue in certain sectors.

Anonimous participant remarked that earlier in 2003, the U.S.A. initiated a case within the World Trade Organization (WTO), challenging the EU's de facto moratorium on the approval of genetically modified (GM) crops. He asked which could be the expected results of this dispute.

G. van Calster answered that he believes that the U.S.A., at the end, will drop the case. New legislation has come into force in the EU. Therefore, the U.S.A. may settle now and then later look at the (new) issues of labelling and traceability. This constitute also one of the EU defence point in the case. Though the U.S.A. might not like the new legislation it is certainly less stringent and more science based. Further it should facilitate the end of the de facto moratorium.

S. Hisano pointed out that Dr. Rudloff mentioned that "communication between assessors with different expertise, of assessors and politicians" is important. He stressed that nobody might be opposed to these ideas. However, when it comes to most of the existing regulatory bodies, including EFSA, it is not necessarily clear what kind of expertise is actually involved in the risk assessment and on what grounds. Moreover, as emphasised above, relevant input from social sciences can improve the openness of the risk assessment process (not just management and communication processes). Although we're likely to consider that only economics (e.g. cost-benefit analysis), or ethics and psychology at most, can contribute to the risk assessment, other fields of (qualitative) social sciences are also useful and necessary especially for *the assessment of risk assessment*, given that science/technology is interwoven with social/economic/political interests and that public concerns are not just a matter of scientifically precise knowledge.

A.D. Hartkamp noted that in her presentation Beate Kettlitz referred to the Eurobarometer 2001: 71% of the consumers don't want biotech food, while 86% indicate that they do not feel sufficiently informed about the technology. She pointed out that she think this is logical, why would you be in favour of something you don't know enough about? The speaker suggests that the public should be involved in the risk analysis process. She pleads to include public concerns (such as ethics, social) and to not dismiss them as irrational. Although *A.D. Hartkamp* expressed her agreement with this principle, she stressed that the public concerns are only legitimate if the public is 'sufficiently' informed. If the public is not sufficiently informed on the technology and the issues at hand the concerns are not always relevant (e.g., for example the concern 'I don't want genes in my food' is not relevant as genes are present in all food, not only GMOs food).

B. Le Gallic highlighted that it was agreed during the workshop that, for risk management and communication purposes, transparency, communication with all stakeholders, trust in institutions, etc. would require new specialised agencies, new regulations, and new forum. Such additional needs have a cost, which should be taken into account in the analysis (whichever should be the social group that will be charged for it), and compared with potential benefits.

R. Rudauskaite noted that as she is working for the Ministry's National Nutrition Centre, her interest was pretty specific and narrow at the beginning of this workshop. After these three days she pointed out that she is completely aware that broad and multidisciplinary work is absolutely necessary at the institutional level though certainly the costs involved might be huge barrier especially for small and not so resourceful countries such as Lithuania.

M. Miraglia stressed that coexistence is certainly one of the crucial issues in the current days. The EU legislation on organic farming (Regulation 2092/91) does not allow the use of GMOs. However seeds with a level of GMOs lower than the threshold do not need to be labelled. Therefore, these might be used even if unwillingly by the farmer. The organic farming regulation does mention the setting of thresholds for unavoidable presence of GMOs and no threshold has been set afterwards. Therefore, it may be advisable that the a EU wide legal binding threshold of GM content in organic products should be established in order to have clear rules for potential legal disputes. Further, she expressed doubts on the current situation where no GMOs (0%) can be used in organic farming while 5% of conventional products (which contain fertilizers and pesticides) is accepted. She concluded asking if consumers are unaware of the presence of conventional products in organic farming or if they instead internalised and accepted this fact.

R. Holzinger commented that 5% tolerance would solve the problem of coexistence. He posed the open question: " Why do bio-Farmers accept 5% contamination with conventionally grown plants, but not with GMOs?". He continued that GMOs farmers should pay when organic farmers have income loss. But do organic farmer pay, when conventional farmer have a contamination (with pests grown on organic fields or with organic plants)? He concluded on this point stating that, according to him, an organic farmer only looses his business, if he agrees with unrealistic production standards, such as 0 % tolerance of GM products.

A. Hozzank replied that consumers expect organic products to be absolutely GM free and they would be disappointed if they discover any GM content. She stated that, in her opinion, this would destroy the market of organic farming.

S. de Vries commented that possibly the times are ripe to concentrate on what we want to achieve (e.g.: reduced use of pesticides, reduced fossil fuel consumption) rather than the methods used to achieve it.

S. Hisano insisted that technology is also an outcome of such processes of social choice, and reproduced and transformed by social activities. It might be needless to say that there have been a lot of claims that strong commercial interests are working behind the research and development of GMOs. Besides, many adverse effects on small family farmers in socio-economic terms are assumed to come about, since this technology is viewed as an important component of the intensive farming system, which many of those groups/individuals calling for stricter regulation of GMOs are opposing. Therefore, if the process of scientific evaluation (i.e. risk assessment) fails to take into consideration these social aspects of technology, and confines its task to what can be handled as a technical, calculable matter, the results of allegedly "sound science"-based evaluation cannot meet the requirement imposed to regain public trust in science and administrations.

H. Valve expressed her concerns on how risks can be made governable. In her opinion the conceptual separation of risk assessment and risk management and the request to keep risk assessment pure from "human influence" seems unrealistic and even dangerous. She observed that the linked issue of science-politics interface in the trade conflict between EU and the US and in the international treaties should be further explored.

B. Le Gallic noted that the workshop showed different views regarding risk assessment, in particular regarding costs associated with the risk of contamination. He continued stating that in absence of historical data, it may be difficult, if not impossible, to provide a definitive answer to this question (which requires to estimate the potential social costs and their probabilities). However, information presented during the workshop may serve as a baseline for scenarios on which policy makers could build upon. He noted that more research and work at the local, national and international level on the economic/social costs/opportunities is needed (e.g. through a comprehensive Costs-Benefits analysis). While such a method, that offers the most comprehensive economic assessment, may be difficult to be applied to every phase of the risk evaluation process (an "ortodox" Cost/Benefit analysis may be often difficult if not unrealistic with risk assessment due to strong uncertainty). On the other hand, economic analysis may at least be conducted for the risk management phase of regulatory impact assessment, and for the risk communication phase.

B. Rudloff stressed that from these 3 days of extremely enriching presentations and discussion she has the feeling that there will be an increasing focus on risk assessment in the coming years. In Europe the major role on this issue should be played by the newly established EFSA. On the other hand, those involved in defining risk assessment are facing enormous problems in their work. Firstly there are immanent scientific problems, for instance a lack of data for next generation of functional food makes risk assessment extremely problematic. When it comes to risk communication it has to be taken into consideration that public "perception" may differ enormously from experts assessment communication. Thus even though experts may see risk as low, the potential risk for GMOs can be perceived as quite high. Perception can be a very large barrier. In addition it is also not easy to find consensus on the definitions on what is a consumer or the public and data can very substantially depending on the reference used.

In closing the workshop *B.Rudloff* expressed her thanks to all the speakers for their valuable presentations as well as the participants for their contribution to the discussion. She also thanked the OECD Co-operative Research Programme: Biological Resource Management for Sustainable Agricultural Systems, for its support as well as the team of local organisers for their indispensable input.

ANNEX 1:

LIST OF PARTICIPANTS

AUSTRALIA

Richard Sisson
Australian Embassy
4 Rue Jean Rey
F-75015. Paris. France
Email: richard.sisson@dfat.gov.au

AUSTRIA

HOZZANK, Alexandra
InfoXgen-Working Group Transparent Food
Königsbrunnerstraße 8
A - 2202 Enzersfeld. Austria
Email: a.hozzank@agrovet.at

BELGIUM

van CALSTER, Geert
Institute of Environmental and Energy Law
Collegium Falconis, K.U. Leuven
Leuven. Belgium
Email: geert.vancalster@law.kuleuven.ac.be

HARMEGNIES, Dimitri
Federal Office for Scientific
Technical and Cultural Affairs (OSTC).
8 Rue de la Science. B-1000 Bruxelles. Belgium
Email: harm@belspo.be

BULGARIA

GENOVA, Gergana
Tzarigradsko Shose' str. 64. BL11.
WH G. AP128.
1784 Sofia. Bulagaria
Email: gerganagenova@abv.bg

NENOV, Asen
26, Golo bardo str., Ent. W
Sofia 1407. Bulgaria
Email: office@adiumdesign.com

CANADA

ALBOVIAS, Anna
Environment Canada.
351 boul. St-joseph. 16th floor.
K1A 0H3. Hull. Canada
Email: anna.albovias@ec.gc.ca

BARTLETT, Diane
6 Norfolk Av.
L6X 2B5. Brampton. Canada
Email: bonbevan@rogers.com

LEE, Stuart
Environment Canada.
8th floor. 351 St.Joseph BLVD.
K1A Gatinau. Quebec. Canada
Email: stuart.lee@ec.gc.ca

DENMARK

MADSEN, Kim-Helleberg
Directorate General Health and Consumer Protection.
Unit D4 – Food Law and biotechnology. European Commission
B-1049 Brussels. Belgium
Email: Kim-Helleberg.Madsen@cec.eu.int

FINLAND

KAUPPILA Jussi
Finnish Environment Institute.
PO Box 140. 00170 Helsinki. Finland
Email: jussi.kauppila@ymparisto.fi

VALVE, Helena
Finnish Environment Institute.
PO Box 140. 00170 Helsinki. Finland
Email: helena.valve@ymparisto.fi
GERMANY

KETTLIZ, Beate
The European Consumers' Organisation (BEUC)
av de Tervueren, 36 bte 4
B - 1040 Bruxelles. Belgium
Email: Food@beuc.org

RUDLOFF, Bettina
The European Institute of Public Administration (EIPA)
O.L. Vrouweplein 22. P.O. Box 1229
NL-6201 BE Maastricht. The Netherlands
Email: b.rudloff@eipa-nl.com

FRANCE

GAL, Jean-Luc
Industrial Property Unit
DG Internal Market, European Commission
B-1049 Brussels. Belgium
Email: jean-luc.gal@cec.eu.int

LE GALLIC, Bertrand
OECD
Fisheries Division. Directorate for Food, Agriculture and Fisheries.
2 Rue Andre-Pascal
F-75775. Paris. Cedex 16. France
Email: bertrand.legallic@oecd.org

HUNGARY

BALÁZS, Ervin
Institute for Environmental Biosafety Research,
Agricultural Biotechnology Center
P.O. Box 411
H-2100 Godollo. Hungary
Email: Balazs@abc.hu

DOMOSKOS, Marton
Gervay Ucta 49.
H-1147. Budapest. Hungary
Email: domokosn.m@mail.datanet.hu

ITALY

CATASTINI, Carlo
Regione Toscana
Via di Novoli 26. I-50127 Firenze. Italy
Email: c.catastini@mail.regione.toscana.it

FOSSI, Fabrizio
Regione Toscana
Via di Novoli 26. I-50127 Firenze. Italy
Email: f.fossi@mail.regione.toscana.it

GALLORI, Enzo
DBAG – Dip.to di Biologia Animale e Genetica
Università degli Studi di Firenze
Via Romana 17,
I-50125 Firenze. Italy
Email: gallori@dbag.unifi.it

GOVI, Daniele
Regione Emilia Romagna
Assessorato Agricoltura, Ambiente e Sviluppo Sostenibile
Viale Silvani 6
I-40122 Bologna. Italy
Email: DGovi@regione.emilia-romagna.it

MARRANI, Daniela
Law firm Portolano Colella Cavallo Prosperetti
Via Catone, 3
I-00192 Roma. Italy
Email: dmarrani@pccp.it

MIRAGLIA, Marina
Istituto Superiore di Sanita'
Via Vitella 33.I-00152. Roma. ItalY
Email: miraglia@iss.it

ZARILLI, Simonetta
UNCTAD
Trade Negotiations and Commercial Diplomacy Branch
Division on International Trade and Commodities
Palais des Nations 8. 14, Av. de la Paix.
1211 Geneva 10. Switzerland
Email: Simonetta.Zarrilli@UNCTAD.org

JAPAN

HISANO, Shuji
Wageningen University
Tarthorst 469
NL-6708 HP Wageningen. The Netherlands
Email: shuji.hisano@wur.nl

INABA, Masakazu
University of Tsukuba
Institute of Biological Sciences
Tsukuba Science City. 305-8572. JAPAN
Email: dj_xis@digital.design.co.jp

LITHUANIA

ANISKEVICIUTE, Girma
Jogailos 12-32
LT-2001. Vilnius. Lithuania
Email: girmaa@delfl.lt

JODINSKAS, Ginataras
Public Agency Nature Heritage Fund.
A. Juozapaviciaus 9. LT 2005. Vilnius. Lithuania
Email: gmo@gamta.lt

RUDAUSKAITE, Ramune
National Nutrition Centre
Kalvariju 153. 2042 Vilnius. Lithuania
Email: ramune@rmc.lt

NETHERLANDS

van DIJK, Piet
DSM Food Specialties
Dept. Regulatory Affairs, 001-0490
P.O. Box 1. NL- 2600 MA Delft. The Netherlands
Email: piet.dijck-van@dsm.com

HARTKAMP, A.D.
Working Group on Biotechnology of the Arable Product Boards
PO BOX 29739.
NL-2502 LS. The Hague. The Netherlands
Email: a.d.hartkamp@hpa.agro.nl

KUIPER, Harry
Food Safety and Health Department (VGV)
State Institute for Quality Control of Agricultural Products (RIKILT-DLO)
Bornsesteeg 45, Postbus 230
NL-6700 AE Wageningen. The Netherlands
Email: Harry.Kuiper@wur.nl

NOTEBORM, H.P.J.M.
Dutch Food and Non-Food Authority (VWA)
Directorate Research and Risk Assessment
Laan van Nieuw-Oost Indië 131
P.O. Box 19506. NL - 2500 CM Den Haag. The Netherlands
Email: h.p.j.m.noteborn@vwa.nl

de VRIES, Sip
Copa-Cogeca
Rue de la Science 23-25
B-1040 Bruxelles. Belgium
Email: s.s.de.vries@hpa.agro.nl

van ROSSUM, Clemens
The Netherlands Health Council
Postbus 16052. NL-2500 CM Den Haag. The Netherlands
Email: clemens.van.rossum@gr.nl

van de WIEL J.A.G.
The Netherlands Health Council
Directorate Research and Risk Assessment
Laan van Nieuw-Oost Indië 131
P.O. Box 19506. NL - 2500 CM Den Haag. The Netherlands
Email: JAG.van.de.Wiel@gr.nl

POLAND

TRACZYK, Iwona
Laboratory of Nutritional Heath Risk Factors.
National Food and Nutrition Institute.
61/63 ul Powsinska. PL-02903. Warsaw. Poland
Email: itraczyk@izz.waw.pl

HORYZA, Joanna
Asnyka 57/21. PL-62 800. Kalisz. Poland
Email: szelmasia@poczta.gazeta.pl

ROMANIA

GRIJA, Adriana Elena
Ministry of Economy and Commerce
Office of State Ownership and Privatization in Industry
152 Calea Victoriei, RO-70034 Bucharest, Sector 1. Romania
Email: adielena@yahoo.com

ISPAS, Iona
National Contact Point for Genomics and Biotechnology for Health.
B-dul Iuliu Maniu 67. BL 67 SC7. Ap. 257.
sector 6. Bucharest. Romania
Email: ioana_isp@hotmail.com

SLOVAKIA

MECAROVA, Zuzana
Palkovicova 3
SK-821 08 Bratislava. Slovakia
Email: zuzana_mecarova@yahoo.com

SZOVICS, Peter
Slovak Agricultural University Nitra
Tr. A. Hlinku 2. SK-94976 Nitra. Slovakia
Email: szovics@yahoo.com

SLOVENIA

Biserka Strel
Ministry of the Environment, Spatial Planning and Energy.
Sector for Biotechnology.
Dunajska 48. 1000 Ljubjana. Slovenia
Email: biserka.strel@gov.si

SWITZERLAND

HOLZINGER, Rainer
Swiss Federal Institute Technology.
Institute of Plant Sciences. ETH Zurich. LFW E 57.1
CH-8092. Zurich. Switzerland
Email: rainer.holzinger@ipw.biol.ethz.ch

THAILAND

MITRCHOB, Chuwit
Royal Thai Embassy. Office of Agricultural Affairs
Avenue Franklin Roosevelt 184
B-1050 Brussels. Belgium
Email: agrithai@skynet.be

UNITED KINGDOM

LEWIS, Chris
General Mills. Harman House.
I George street. UB8 1QQ. Uxbridge. UK
Email: chris.lewis@genmills.com

MACKENZIE, Ruth
Foundation for Environmental Law and Development (FIELD).
52-53 Russell Square, London WC1B 4HP. UK
Email: rmackenzie@mail.field.org.uk

OLIVER, Simon
Old El Paso Foods BV.
Postbus 1091. NL-1380 AC Weesp. The Netherlands
Email: n.a.

WILLIAMS, Katherine
AgraFood Biotech – Agra Europe
80 Calverley Road. Tunbridge Wells
Kent. TN1 2UN. UK
Email: katherine.williams@informa.com

UKRAINE

PASISHNYK, Natalia
Institute of Forest Economics
Albert-Ludwigs-University Freiburg
Tennenbacher Str. 4
79106 Freiburg. Germany
Permanent address:
81100 s.Malechkovychi
Pustomytivsky r-n Lvivska obl.
UKRAINE
Email: nataliya.pasishnyk@neptun.uni-freiburg.de

USA

MCHUGHEN, Alan
FACN, University of California at Riverside
Riverside, CA 92521-0124
USA
Email: alanmc@citrus.ucr.edu

PLOEHN, Jerry
Biotechnology Working Group
The National Corn Growers Association (NCGA)
85703 300th Street
Alpha, MN 56111. USA
Email: jploehn@direcway.com

SCHEPERS, James
USDA-ARS-NPA, Department of Agronomy
University of Nebraska
113 Keim Hall
Lincoln, NE 68583-0915. USA
Email: jschepers1@unl.edu

ANNEX 2

ORGANISING COMMITTEE

OECD Scientific co-ordinators
James Schepers
Ervin Balazs
Local Organising Committee
Ruggero Lala, AMSU
Sabine Nibeling, AMSU
Bettina Rudloff, EIPA
Winny Curfs, EIPA
Workshop Secretariat
Ruggero Lala, AMSU

OECD PUBLICATIONS, 2, rue André-Pascal, 75775 PARIS CEDEX 16
PRINTED IN FRANCE
(51 2004 07 1 P) ISBN 92-64-10877-7 – No. 53657 2004